IAENG TRANSACTIONS ON ENGINEERING SCIENCES

Special Issue for the International Association of Engineers Conferences 2016

Editors

Sio-Iong Ao
International Association of Engineers, Hong Kong

Alan Hoi-shou Chan
City University of Hong Kong, Hong Kong

Hideki Katagiri
Kanagawa University, Japan

World Scientific

NEW JERSEY • LONDON • SINGAPORE • BEIJING • SHANGHAI • HONG KONG • TAIPEI • CHENNAI • TOKYO

Published by

World Scientific Publishing Co. Pte. Ltd.

5 Toh Tuck Link, Singapore 596224

USA office: 27 Warren Street, Suite 401-402, Hackensack, NJ 07601

UK office: 57 Shelton Street, Covent Garden, London WC2H 9HE

Library of Congress Cataloging-in-Publication Data

Names: Ao, Sio-Iong, editor. | Chan, Alan H. S., editor. | Katagiri, Hideki, editor. | International Association of Engineers, sponsoring body. | International Multi-Conference of Engineers and Computer Scientists (2016 : Hong Kong, China), author. | World Congress on Engineering (2016 : London, England), author.

Title: Special issue for the International Association of Engineers conferences 2016 / edited by: Sio-Iong Ao (International Association of Engineers, Hong Kong), Alan Hoi-shou Chan (City University of Hong Kong, Hong Kong), Hideki Katagiri (Hiroshima University, Japan).

Description: New Jersey : World Scientific, [2017] | Series: IAENG transactions on engineering sciences | Contains 21 revised papers from Hong Kong, March 16–18, 2016, the International MultiConference of Engineers and Computer Scientists (IMECS 2016), and London, UK, June 29–July 1, 2016, under the World Congress on Engineering (WCE 2016). | Includes bibliographical references.

Identifiers: LCCN 2017017198 | ISBN 9789813226197 (hc : alk. paper)

Subjects: LCSH: Engineering--Congresses. | Computer science--Congresses.

Classification: LCC TA5 .S6642 2017 | DDC 620--dc23

LC record available at https://lccn.loc.gov/2017017198

British Library Cataloguing-in-Publication Data

A catalogue record for this book is available from the British Library.

Desk Editor: V. Vishnu Mohan

Typeset by Stallion Press
Email: enquiries@stallionpress.com

Printed in Singapore

PREFACE

Two large international, conferences on Advances in Engineering Sciences were held in Hong Kong, March 16–18, 2016, under the International MultiConference of Engineers and Computer Scientists (IMECS 2016), and in London, U.K., 29 June–1 July, 2016, under the World Congress on Engineering 2016 (WCE 2016) respectively. The IMECS 2016 and WCE 2016 are organized by the International Association of Engineers (IAENG). IAENG is a non-profit international association for the engineers and the computer scientists, which was founded originally in 1968 and has been undergoing rapid expansions in recent few years. The conferences serve as good platforms for the engineering community to meet with each other and to exchange innovative research ideas. The conferences have also struck a balance between theoretical and application development. The conference committees have been formed with over six hundred committee members who are mainly research center heads, faculty deans, department heads, professors, and research scientists from over 30 countries with committee member details available at:

http://www.iaeng.org/IMECS2016/committee.html
http://www.iaeng.org/WCE2016/committee.html

The conferences are truly international meetings with a high level of participation from many countries. The response that we have received for the congresses is excellent. There have been more than five hundred manuscript submissions for the IMECS 2016. All submitted papers have gone through the peer review process and the overall acceptance rate is 50.39%. For the WCE 2016, there have been more than seven hundred manuscript submissions. All submitted papers have gone through the peer review process and the overall acceptance rate is 50.82%.

This volume contains 21 revised and extended research articles written by prominent researchers participating in the conferences. Topics covered include engineering mathematics, computer science, electrical engineering, manufacturing engineering, industrial engineering, and industrial applications. The book offers the state of art of tremendous advances in engineering sciences and also serves as an excellent reference work for researchers and graduate students working with/on engineering sciences.

Sio-Iong Ao
Alan Hoi-shou Chan
Hideki Katagiri

CONTENTS

AN APPLICATION OF THE CHANGE-POINT ESTIMATION FOR THE POISSON DISTRIBUTION

D. GHORBANZADEH, Ph. DURAND* and LUAN JAUPI

Conservatoire National des Arts et Métiers, Département Mathématiques, 292, rue Saint martin, 75141, Paris, France

**philippe.durand@cnam.fr*

Inference is considered about the point in a sequence of random variables at which the probability distribution changes. In particular, we consider the case a Poisson distribution with changing in parameter.

Keywords: Change-point; likelihood estimator; Poisson distribution.

1. Introduction

In this paper presented at WCE in London,[1] we are interested by study the estimation of the change-point. The applications of change-point analysis are huge: we find relevant literature in many fields including: Biology, Chemistry, Environmental Sciences and Climate Change, Engineering, Econometrics, Medicine, Behavioral Sciences and also in Political Science, Finance, Image Analysis and Security.

When the observations reveal a phenomenon of change point, it is useful to determine firstly the relevant models and also to specify where the break occurs. The aim of this work is to present a theoretical point of view different types of point of change and to demonstrate the use of these models in an application. The application concerns Poisson observations.

we apply the results to the data in Refs. 2 and 3, the data represent the number of haemolytic uraemic syndrome (HUS) cases in Birmingham and Newcastle from 1970 to 1989. We also apply our results to data in Ref. 4, the data represent the number of earthquakes in Northern California from 1984 to 2011.

2. Statistical model

The observed data are a sequence of random variables $X_1, \ldots, X_n$ that may change law after the first k observations. The time k (called the

time of change-point), strictly between 1 and n such that the observations $X_1, \ldots, X_k$ are independent and identically distributed according to a distribution $\mathbb{P}_{\theta_1}$ and observations $X_{k+1}, \ldots, X_n$ are independent and identically distributed according to a distribution $\mathbb{P}_{\theta_2}$.
The model parameters are estimated from the likelihood function (the maximum likelihood method). The parameter estimates θ_1 and θ_2 will depend on the time k. In other words, each time k we have a probability, thus estimate the parameter k, we consider the function likelihood maximum.

3. Model for the Poisson observations

We consider the following model:

$$\begin{cases} X_1, \ldots, X_k \sim \mathcal{P}(\lambda) \\ X_{k+1}, \ldots, X_n \sim \mathcal{P}(c\lambda) \end{cases} \tag{1}$$

where $\lambda > 0$ and $c > 0$.
In the following, we will note

$$S_k = \sum_{i=1}^{k} X_i \, , \; S_k^\star = \sum_{i=k+1}^{n} X_i \, , \; \overline{X}_k = \frac{S_k}{k} \, , \; \overline{X_k^\star} = \frac{S_k^\star}{n-k} \, .$$

The likelihood function is given by

$$\begin{aligned} L(x, \lambda, c, k) &= \prod_{i=1}^{k} e^{-\lambda} \frac{\lambda^{x_i}}{x_i!} \prod_{i=k+1}^{n} e^{-c\lambda} \frac{(c\lambda)^{x_i}}{x_i!} \\ &= \left(\prod_{i=1}^{n} \frac{1}{x_i!} \right) e^{-k\lambda-(n-k)c\lambda} \, \lambda^{s_k+s_k^\star} \, c^{s_k^\star} \end{aligned} \tag{2}$$

The log-Likelihood function is given by

$$\begin{aligned} \ell(x, \lambda, c, k) &= \log\big(L(x, \lambda, c, k)\big) \\ &= \log \left(\prod_{i=1}^{n} \frac{1}{x_i!} \right) - k\lambda - (n-k)c\lambda + (s_k + s_k^\star) \log\lambda + s_k^\star \log c \end{aligned} \tag{3}$$

The maximum likelihood estimators $\hat{\lambda}$ and $\hat{c}$ is then that value of λ and c that maximises ℓ defined in equation (3). They are determined by taking the partial derivative of ℓ with respect to λ and c and setting it to zero.

We have

$$\begin{cases} \dfrac{\partial}{\partial \lambda}\ell(x,\lambda,c,k) = -k-(n-k)c+\dfrac{s_k+s_k^\star}{\lambda} \\ \dfrac{\partial}{\partial c}\ell(x,\lambda,c,k) = -(n-k)\lambda+\dfrac{s_k^\star}{c} \end{cases} \tag{4}$$

From equation (4), we deduce the maximum likelihood estimators $\hat{\lambda}$ and $\hat{c}$,

$$\hat{\lambda} = \overline{X}_k \ , \ \hat{c} = \frac{\overline{X_k^\star}}{\overline{X}_k} \tag{5}$$

By injecting $\hat{\lambda}$ and $\hat{c}$ in equation (3), we obtain

$$R(k) = \ell(x,\hat{\lambda},\hat{c},k) = \log\left(\prod_{i=1}^{n}\frac{1}{x_i!}\right) - s_n + s_k\log(\overline{x}_k) + s_k^\star\log(\overline{x_k^\star}) \tag{6}$$

where $s_n = s_k + s_k^\star$. The maximum likelihood estimators $\hat{k}$ is defined by,

$$\hat{k} = \max_{1<k<n-1} R(k). \tag{7}$$

4. Simulation results

In order to study the performance of the method, we simulated a sample of size 500 and the values set $\lambda = 5$, $c = 2$ and $k = 150$. Figure 1 show the results obtained.

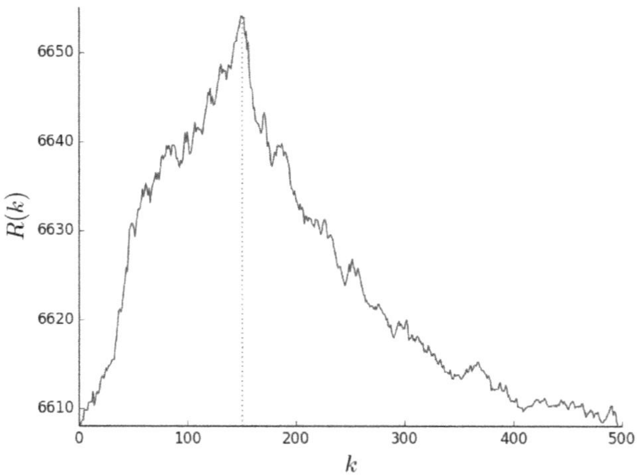

Fig. 1. Graph of $R(k)$, with the values estimator $\hat{\lambda} = 5.198675$, $\hat{c} = 1.458383$ and $\hat{k} = 151$.

5. Applications

5.1. *The number of HUS*

In this section we apply the previous results to the data used by Refs. 2 and 3. The data represent the number of cases of diarrhoea-associated haemolytic uraemic syndrome (HUS) for the years 1970–1989, in two cities Birmingham and Newcastle within the United Kingdom.

Table 1 displays the annual number of cases of HUS collected in Birmingham and Newcastle of England, respectively, from 1970 to 1989.

Table 1. The annual number of cases of HUS.

Year	Count at Birmingham	Count at Newcastle	Year	Count at Birmingham	Count at Newcastle
1970	1	6	1980	1	4
1971	5	1	1981	7	0
1972	3	0	1982	11	4
1973	2	0	1983	4	3
1974	2	2	**1984**	**7**	**3**
1975	1	0	1985	10	13
1976	0	1	1986	16	14
1977	0	8	1987	16	8
1978	2	4	1988	9	9
1979	1	1	1989	15	19

Figures 2 and 3 show $R(k)$, for the Birmingham data and Newcastle data, based on k, the maximum is reached at the point $\hat{k} = 15$ (year 1984).

5.2. *The number of earthquakes*

In this section we apply the results to the data used by Ref. 4. This data includes high-precision hypocenter locations for 513474 earthquakes recorded between January 1984–December 2011 by stations of the Northern California Seismic System (NCSS). For our application, we chose the number of earthquakes with magnitude ≥ 3.

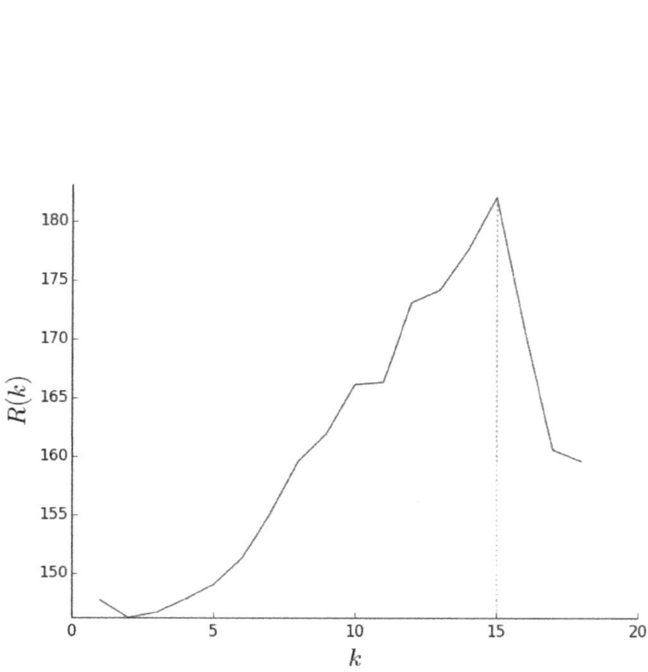

Fig. 2. Graph of $R(k)$ for the Birmingham data, with $\hat{\lambda} = 2.066667$ and $\hat{c} = 6.096774$.

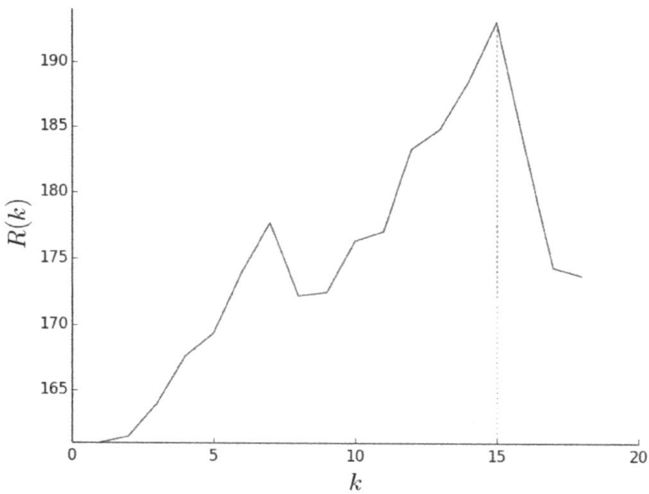

Fig. 3. Graph of $R(k)$ for the Newcastle data, with $\hat{\lambda} = 2.466667$ and $\hat{c} = 5.108108$.

Table 2 displays the number of earthquakes in Northern California with magnitude ≥ 3 from 1984 to 2011.

Table 2. The number of earthquakes in Northern California.

Year	Number	Year	Number	Year	Number	Year	Number
1984	539	1991	337	1998	375	**2005**	**194**
1985	329	1992	555	1999	352	2006	172
1986	682	1993	489	2000	264	2007	186
1987	298	1994	450	2001	311	2008	197
1988	270	1995	335	2002	193	2009	178
1989	460	1996	283	2003	422	2010	228
1990	343	1997	374	2004	353	2011	181

Figure 4 shows $R(k)$, for the number of earthquakes in Northern California with magnitude ≥ 3, based on k, the maximum is reached at the point $\hat{k} = 22$ (year 2005).

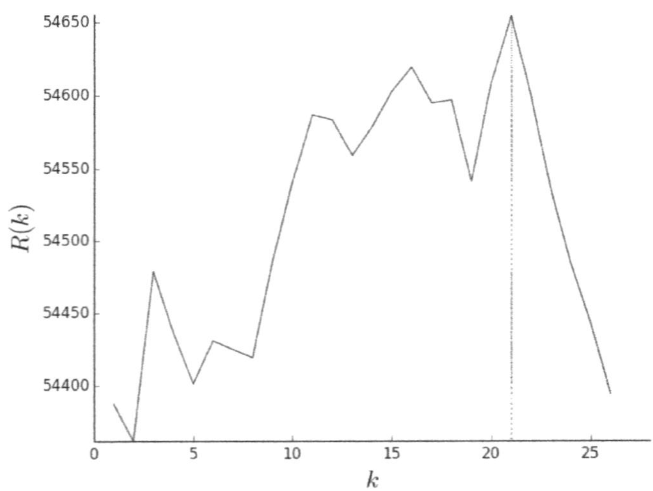

Fig. 4. Graph of $R(k)$ for the Northern California data, with $\hat{\lambda} = 381.619048$ and $\hat{c} = 0.500125$.

6. Conclusion

The method used in this work is very simple to implement and particularly easy to program. In the application part, we obtained the same results as Refs. 2 and 3 who used another change point detection method.

References

1. D. Ghorbanzadeh, P. Durand and L. Jaupi (2016). "Problem of Change-Point Detection for the Poisson Observations", Lecture Notes in Engineering and Computer Science: Proceedings of The World Congress on Engineering 2016, 29 June-1 July, 2016, London, U.K., pp. 20-22.
2. R. Henderson and J. N. S. Matthews (1993). An investigation of change-points in the annual number of cases of haemolytic uraemic syndrome. Appl. Statist., 42(3), 461-471.
3. R. W. West and R. T. Ogden (1997). Continuous-time estimation of a change-point in a Poisson process. J. Statist. Comput. Simul., 56, 293-302.
4. F. Waldhauser and D. P. Schaff (2008). Large-scale relocation of two decades of Northern California seismicity using cross-correlation and double-difference methods, J. Geophys. Res., 113, B08311, doi:10.1029/2007JB00 5479.

MIXED INTERAL EQUATIONS AND THEIR APPLICATION TO MECHANICS

A. V. MANZHIROV

Ishlinsky Institute for Problems in Mechanics of the Russian Academy of Sciences,
Vernadsky Ave 101 Bldg 1, Moscow, 119526, Russia
Bauman Moscow State Technical University,
2nd Baumanskaya Str 5 1, Moscow, 105005, Russia
National Research Nuclear University MEPhI,
Kashirskoye shosse 31, Moscow, 115409, Russia
Moscow Technological University,
Vernadsky Ave 78, Moscow, 119454, Russia
manzh@inbox.ru
www.ipmnet.ru

Integral equations with several different operators often arise in various fields of natural sciences. We call them mixed integral equations. One can find an integral equation with one Volterra and one compact operator in tribology and an equation with two Volterra operators and one compact operator in contact mechanics. Such equations are very specific and it is necessary to develop new approach for their solution. This paper deals with a new projection method for analytic solution of these equations. We consider the general case and, in particular,the contact problem for a viscoelastic foundation with a nonuniform coating. We write out the solution which can't be obtained by any other known method with sufficient accuracy.

Keywords: Contact Mechanics; Integral Equation; Integral Operator; Tribology; Volterra Operator.

Introduction

Integral equations with several different operators arose in the 70th–80th of XX century as a result of mathematical modeling of wear process (see, e.g., [1–2]) and contact interaction (see, e.g., [1–10]) for viscoelastic foundations with thin coatings. First attempts to obtain the solutions of these equations allowed to reduce them to solving of systems of infinite systems of Volterra integral equations which have no still analytical and numerical research methods. Moreover, contact problems for solids with nonuniform coatings and coatings with rough surface have posed another mathematical problem regarding adequate methods for equations with rapidly oscillating functions in initial data. We consider both problems and construct new efficient projection method as well as demonstrate its application to contact problem in mechanics (see also [11–12]).

1. Integral Equation with a Known Right-Hand Side

Consider a mixed multi-dimensional equation with integral operators of Volterra and Schmidt types

$$\begin{aligned} &\sigma(t)(\mathcal{I}-\mathcal{V}_1)y(\mathbf{x},t)+(\mathcal{I}-\mathcal{V}_2)\mathcal{S}y(\mathbf{x},t)=\frac{f(\mathbf{x},t)}{h(\mathbf{x})},\\ &\mathcal{S}y(\mathbf{x},t)=\int_\Omega S(\mathbf{x},\boldsymbol{\xi})y(\boldsymbol{\xi},t)\,d\Omega_\xi,\quad S(\mathbf{x},\boldsymbol{\xi})=\frac{F(\mathbf{x},\boldsymbol{\xi})}{h(\mathbf{x})},\\ &\mathcal{V}_p y(\mathbf{x},t)=\int_{\tau_0}^t V_p(t,\tau)y(\mathbf{x},\tau)\,d\tau,\quad \mathbf{x}\in\Omega,\quad \tau_0\le t\le T. \end{aligned} \tag{1}$$

In this section, we consider some general questions of the theory of mixed equations.

Let the right-hand side $f(\mathbf{x},t)/h(\mathbf{x})$ of equation (1) be known. It is required to find the function $y(\mathbf{x},t)$. Here, $f(\mathbf{x},t)$ and $y(\mathbf{x},t)$ are continuous functions of $t\in[1,T]$ with values in $L_2(\Omega)$; $\sigma(t)$ is a given positive continuous function; $h(\mathbf{x})>0$ is a given function of class $L_2(\Omega)$; $F(\mathbf{x},\boldsymbol{\xi})$ is a symmetric positive definite Fredholm kernel; $\mathcal{V}_1$ and $\mathcal{V}_2$ are Volterra operators; $\mathcal{S}$ is a Schmidt operator.

Let us transform the equation with the Schmidt operator to an equation with a Hilbert–Schmidt operator. To this end, we multiply (1) by $\sqrt{h(\mathbf{x})}$ and change the variables as follows

$$q(\mathbf{x},t)=\sqrt{h(\mathbf{x})}y(\mathbf{x},t),\quad F^h(\mathbf{x},\boldsymbol{\xi})=\frac{S(\mathbf{x},\boldsymbol{\xi})\sqrt{h(\mathbf{x})}}{\sqrt{h(\boldsymbol{\xi})}}=\frac{F(\mathbf{x},\boldsymbol{\xi})}{\sqrt{h(\mathbf{x})h(\boldsymbol{\xi})}}. \tag{2}$$

Then

$$\begin{aligned} &\sigma(t)(\mathcal{I}-\mathcal{V}_1)q(\mathbf{x},t)+(\mathcal{I}-\mathcal{V}_2)\mathcal{F}^h q(\mathbf{x},t)=\frac{f(\mathbf{x},t)}{\sqrt{h(\mathbf{x})}},\\ &\mathcal{F}^h q(\mathbf{x},t)=\int_\Omega F^h(\mathbf{x},\boldsymbol{\xi})q(\boldsymbol{\xi},t)\,d\Omega_\xi,\quad \mathbf{x}\in\Omega,\quad \tau_0\le t\le T. \end{aligned} \tag{3}$$

where $q(\mathbf{x},t)$ and $f(\mathbf{x},t)/\sqrt{h(\mathbf{x})}$ are continuous functions of $t\in[\tau_0,T]$ with values in the Hilbert space $L_2(\Omega)$; $\mathcal{F}^h$ is a Hilbert–Schmidt operator, and the other functions have been specified above.

Suppose that the right-hand side of equation (3) is known and we have to find the function $q(\mathbf{x},t)$.

Let us seek a solution of the mixed equation (3) in the form of a series

$$q(\mathbf{x},t) = \sum_{k=1}^{\infty} q_k(t)\varphi_k^h(\mathbf{x}), \tag{4}$$

where $\varphi_k^h(\mathbf{x})$ are eigenfunctions of the operator $\mathcal{F}^h$ corresponding to eigenvalues $\mu_k^h > 0$, i.e.,

$$\mathcal{F}^h\varphi_k^h(\mathbf{x})\,d\xi = \mu_k^h\varphi_k^h(\mathbf{x}), \quad k = 1,2,\ldots \tag{5}$$

The representation (4) is possible, since the system of eigenfunctions of the operator $\mathcal{F}^h$ forms a basis in $L_2(\Omega)$.

Let us construct the functions of the basis in the form

$$\varphi_k^h(\mathbf{x}) = \frac{\Phi_k^h(\mathbf{x})}{\sqrt{h(\mathbf{x})}}, \quad k = 1,2,\ldots \tag{6}$$

with explicit dependence on the function $h(\mathbf{x})$, where

$$\int_\Omega \varphi_i^h(\boldsymbol{\xi})\varphi_j^h(\boldsymbol{\xi})\,d\Omega_\xi = \int_\Omega \frac{\Phi_i^h(\boldsymbol{\xi})\Phi_j^h(\boldsymbol{\xi})}{h(\boldsymbol{\xi})}\,d\Omega_\xi = \delta_{ij} = \begin{cases} 1, & \text{for } i = j, \\ 0, & \text{for } i \neq j. \end{cases} \tag{7}$$

In order to construct such eigenfunctions, we first construct a basis $p_n^h(\mathbf{x})$ in $L_2(\Omega)$ for which

$$\int_\Omega p_i^h(\boldsymbol{\xi})p_j^h(\boldsymbol{\xi})\,d\Omega_\xi = \delta_{ij}, \quad p_n^h(\mathbf{x}) = \frac{P_n^h(\mathbf{x})}{\sqrt{h(\mathbf{x})}}, n = 1,2,\ldots \tag{8}$$

Such a basis can be constructed by the formulas

$$\begin{gathered} P_1^h(\mathbf{x}) = \frac{f_1(\mathbf{x})}{\sqrt{H_{11}}}, \quad \Delta_0 = 1, \quad \Delta_1 = H_{11}, \\ \Delta_n = \begin{vmatrix} H_{11}\,H_{12}\ldots H_{1n} \\ H_{21}\,H_{22}\ldots H_{2n} \\ \vdots \quad \vdots \quad \ddots \quad \vdots \\ H_{n1}H_{n2}\ldots H_{nn} \end{vmatrix}, \quad H_{ij} = \int_\Omega \frac{f_i(\boldsymbol{\xi})f_j(\boldsymbol{\xi})}{h(\boldsymbol{\xi})}\,d\Omega_\xi, \\ P_n^h(\mathbf{x}) = \frac{1}{\sqrt{\Delta_{n-1}\Delta_n}} = \begin{vmatrix} H_{11} & H_{12} & \ldots & H_{1n} \\ H_{21} & H_{22} & \ldots & H_{2n} \\ \vdots & \vdots & \ddots & \vdots \\ f_1(\mathbf{x}), & f_2(\mathbf{x}) & \ldots & f_n(\mathbf{x}) \end{vmatrix}, \end{gathered} \tag{9}$$

where $f_i(\mathbf{x})$ is an arbitrary complete system of linearly independent function in $L_2(\Omega)$.

Let us represent the kth eigenfunction in the form of a series with respect to the basis $p_i^h(\mathbf{x})$ of $L_2(\Omega)$. We have

$$\varphi_k^h(\mathbf{x}) = \sum_{i=1}^{\infty} \varphi_{i(k)}^h p_i^h(\mathbf{x}), \quad p_i^h(\mathbf{x}) = \frac{P_i^h(\mathbf{x})}{\sqrt{h(\mathbf{x})}}, \quad \Phi_k^h(\mathbf{x}) = \sum_{i=1}^{\infty} \varphi_{i(k)}^h P_i^h(\mathbf{x}). \tag{10}$$

The Hilbert–Schmidt kernel $F^h(\mathbf{x}, \boldsymbol{\xi})$ can be expanded into double series with respect to the chosen basis

$$\begin{aligned} F^h(\mathbf{x}, \boldsymbol{\xi}) &= \sum_{m=1}^{\infty} \sum_{n=1}^{\infty} F_{mn}^h p_m^h(\mathbf{x}) p_n^h(\boldsymbol{\xi}), \\ F_{mn}^h &= \int_{\Omega}\!\int_{\Omega} F^h(\mathbf{x}, \boldsymbol{\xi}) p_m^h(\mathbf{x}) p_n^h(\boldsymbol{\xi})\, d\Omega_x\, d\Omega_\xi, \quad F_{mn}^h = F_{nm}^h. \end{aligned} \tag{11}$$

Substituting (10) and (11) into (5), we obtain an infinite system of linear algebraic equations for the determination of the eigenvalues and the eigenfunction expansion coefficients. This system has a symmetric matrix and can be written as follows

$$\sum_{n=1}^{\infty} F_{mn}^h \varphi_{n(k)}^h = \mu_k^h \varphi_{m(k)}^h, \quad m = 1, 2, \ldots \tag{12}$$

In order to calculate approximations for N eigenvalues and eigenfunctions of the Hilbert–Schmidt operator $\mathcal{F}^h$, it is necessary to find the eigenvalues and orthonormal eigenvectors of the matrix

$$[F_{NN}^h] = \begin{pmatrix} F_{11}^h & \cdots & F_{1N}^h \\ \vdots & \vdots & \vdots \\ F_{1N}^h & \cdots & F_{NN}^h \end{pmatrix}. \tag{13}$$

The eigenvalues of the matrix (13) give approximations for the first N eigenvalues of the Hilbert–Schmidt operator, and the components of its orthonormal eigenvectors give approximate values of the expansion coefficients for the first N eigenfunctions of that series with N orthonormal functions of the basis.

Let us write the right-hand side of equation (3) in the form

$$\begin{aligned} \frac{f(\mathbf{x}, t)}{\sqrt{h(\mathbf{x})}} &= \sum_{k=1}^{\infty} f_k^h(t) \varphi_k^h(\mathbf{x}) = \sum_{k=1}^{\infty} f_k^h(t) \frac{\Phi_k^h(\mathbf{x})}{\sqrt{h(\mathbf{x})}}, \\ f_k^h(t) &= \int_{\Omega} \frac{f(\boldsymbol{\xi}, t)}{\sqrt{h(\boldsymbol{\xi})}} \varphi_k^h(\boldsymbol{\xi})\, d\Omega_\xi = \int_{-1}^{1} \frac{f(\mathbf{x}, t)}{h(\mathbf{x})} \Phi_k^h(\mathbf{x})\, dx. \end{aligned} \tag{14}$$

Substituting (4), (14) into (3) and taking into account (5), we obtain the following sequence of Volterra equations for the unknown functions $q_k(t)$:

$$
\begin{gathered}
(\mathcal{I}-\mathcal{V}_k^h)q_k(t)=\delta_k^h(t), \quad \delta_k^h(t)=\frac{f_k^h(t)}{\sigma(t)+\mu_k^h}, \\
\mathcal{V}_k^h=\frac{\sigma(t)\mathcal{V}_1+\mu_k^h\mathcal{V}_2}{\sigma(t)+\mu_k^h}, \quad \mathcal{V}_k^h f(t)=\int_{\tau_0}^{t} V_k^h(t,\tau)f(\tau)\,d\tau, \\
V_k^h=\frac{\sigma(t)V_1(t,\tau)+\mu_k^h V_2(t,\tau)}{\sigma(t)+\mu_k^h}, \quad k=1,2,\ldots
\end{gathered}
\tag{15}
$$

where all operators $\mathcal{V}_k^h$ are of Volterra type, just as the operators $\mathcal{V}_1$ and $\mathcal{V}_2$, since $\mu_k^h\to 0$ as $k\to\infty$.

A solution of the sequence of Volterra equations (15) can be constructed by any known analytical and numerical method. This solution can be written in the form

$$
\begin{gathered}
q_k(t)=(\mathcal{I}+\mathcal{R}_k^h)\,\delta_k^h(t), \quad (\mathcal{I}-\mathcal{V}_k^h)^{-1}=(\mathcal{I}+\mathcal{R}_k^h), \\
\mathcal{R}_k^h f(t)=\int_{\tau_0}^{t} R_k^h(t,\tau)f(\tau)\,d\tau,
\end{gathered}
\tag{16}
$$

where $\mathcal{R}_k^h$ is the resolvent operator for $\mathcal{V}_k^h$, and $R_k(t,\tau)$ is the resolvent of the kernel $V_k^h(t,\tau)$. The series (13) converges in $L_2(\Omega)$ uniformly in $t\in[\tau_0,t]$, and its sum is a continuous function of t with values in $L_2(\Omega)$.

Finally, taking into account (2), (4), and (6), (16), we find that

$$
y(\mathbf{x},t)=\frac{1}{h(\mathbf{x})}\sum_{k=1}^{\infty}(\mathcal{I}+\mathcal{R}_k^h)\delta_k^h(t)\Phi_k^h(\mathbf{x}). \tag{17}
$$

Note that the function $h(\mathbf{x})$ enters the solution (17) in explicit form, which allows us to solve equation (1) with high accuracy, even for a rapidly changing function $h(\mathbf{x})$.

2. Integral Equation with Auxiliary Conditions

Consider equation (1) with the right-hand side $f(\mathbf{x},t)=\sum_{i=1}^{N}\alpha_i(t)f_i(\mathbf{x})-g(\mathbf{x},t)$ and N auxiliary integral conditions on the unknown function $y(x,t)$, where $\mathbf{x}\in\Omega$ and $\tau_0\le t\le T$.

The problem is to find a solution of the integral equation

$$
\sigma(t)(\mathcal{I}-\mathcal{V}_1)y(\mathbf{x},t)+(\mathcal{I}-\mathcal{V}_2)\mathcal{S}y(\mathbf{x},t)=\sum_{i=1}^{N}\alpha_i(t)\frac{f_i(\mathbf{x})}{h(\mathbf{x})}-\frac{g(\mathbf{x},t)}{h(\mathbf{x})}, \tag{18}
$$

with the auxiliary conditions

$$\int_\Omega y(\boldsymbol{\xi},t) f_i(\boldsymbol{\xi}) d\Omega_\xi = M_i(t), \quad i=1,\ldots,N, \tag{19}$$

regarding $y(\mathbf{x},t)$ and $\alpha_1(t),\ldots,\alpha_N(t)$ as unknown functions. All other functions in (18) are assumed given, and $g(\mathbf{x},t)$ is a continuous function of t with values in $L_2(\Omega)$; $f_i(\boldsymbol{\xi})$ is a system of N linearly independent functions in $L_2(\Omega)$.

Let us transform the equation with the Schmidt operator to an equation with a Hilbert–Schmidt operator by changing the variables as in (2). Then, equation (18) and the auxiliary conditions (19) become

$$\sigma(t)(\mathcal{I}-\mathcal{V}_1)q(\mathbf{x},t) + (\mathcal{I}-\mathcal{V}_2)\mathcal{F}^h q(\mathbf{x},t) = \sum_{i=1}^N \alpha_i(t)\frac{f_i(\mathbf{x})}{\sqrt{h(\mathbf{x})}} - \frac{g(\mathbf{x},t)}{\sqrt{h(\mathbf{x})}}, \tag{20}$$

$$\int_\Omega q(\boldsymbol{\xi},t)\frac{f_i(\boldsymbol{\xi})}{\sqrt{h(\boldsymbol{\xi})}}\, d\Omega_\xi = M_i(t), \quad i=1,\ldots,N. \tag{21}$$

In order to construct a solution of the mixed integral equation (18) with the auxiliary conditions (19), we construct a special basis in $L_2(\Omega)$ with explicit dependence on the function $1/\sqrt{h(\mathbf{x})}$. To this end, we complement the system of N linearly independent functions $f_i(\mathbf{x})$, so as to obtain a complete system in $L_2(\Omega)$, and then use formulas (8) and (9). As a result, we obtain a basis $p_n^h(\mathbf{x})$ in $L_2(\Omega)$ for which (in view of (8)and (9)) the following expansion holds

$$p_i^h(\mathbf{x}) = \frac{P_i^h(\mathbf{x})}{\sqrt{h(\mathbf{x})}} = \sum_{k=1}^i a_{ik}\frac{f_k(\mathbf{x})}{\sqrt{h(\mathbf{x})}}, \quad i=1,\ldots,N. \tag{22}$$

Resolving the system of algebraic equations (22), we obtain

$$\frac{f_i(\mathbf{x})}{\sqrt{h(\mathbf{x})}} = \sum_{k=1}^i b_{ik} p_k^h(\mathbf{x}), \quad i=1,\ldots,N, \tag{23}$$

the matrix of system (23) being the inverse of the matrix corresponding to system (22).

Let us represent the Hilbert space $L_2(\Omega)$ as a direct sum of its orthogonal subspaces

$$L_2(\Omega) = L_2^\circ(\Omega) \oplus L_2^*(\Omega), \tag{24}$$

where $L_2^\circ(\Omega)$ is the Euclidean space with the basis $p_1(\mathbf{x}),\ldots,p_N(\mathbf{x})$, and $L_2^*(\Omega)$ is the Hilbert space with the basis $\{p_k(\mathbf{x})\}$ $(k=N+1,N+2,\ldots)$.

Note that any continuous function of t with values in $L_2(\Omega)$ can be represented as a sum of continuous functions of t with values in $L_2^\circ(\Omega)$ and $L_2^*(\Omega)$. Let us write such a representation for the integrand

$$q(\mathbf{x},t)=q^\circ(\mathbf{x},t)+q^*(\mathbf{x},t),\quad q^\circ(\mathbf{x},t)=\sum_{n=1}^{N} q_n^\circ(t)p_n^h(\mathbf{x}). \tag{25}$$

Using the auxiliary conditions (21), together with (23) and (25), we obtain the following system of equations

$$\sum_{k=1}^{i} b_{ik}q_n^\circ(t) = M_i(t),\quad i=1,\ldots,N. \tag{26}$$

The solution of this system determines the coefficients of the first term in the expansion (25) of $q^\circ(\mathbf{x},t)$:

$$q_i^\circ(t) = \sum_{k=1}^{i} a_{ik}M_k(t),\quad i=1,\ldots,N. \tag{27}$$

In view of (23), the right-hand side of the equation can be written in the form

$$\begin{aligned}
&\frac{f(\mathbf{x},t)}{\sqrt{h(\mathbf{x})}} = \sum_{i=1}^{N}\alpha_i(t)\frac{f_i(\mathbf{x})}{\sqrt{h(\mathbf{x})}} - \frac{g(\mathbf{x},t)}{\sqrt{h(\mathbf{x})}} = f_h^\circ(\mathbf{x},t)+f_h^*(\mathbf{x},t),\\
&f_h^\circ(\mathbf{x},t) = \sum_{k=1}^{N}\sum_{i=k}^{N}[\alpha_i(t)b_{ik} - g_k^{h\circ}(t)]p_k^h(\mathbf{x}),\quad f_h^*(\mathbf{x},t) = -g_h^*(\mathbf{x},t),\\
&\frac{g(\mathbf{x},t)}{\sqrt{h(\mathbf{x})}} = g_h^\circ(\mathbf{x},t)+g_h^*(\mathbf{x},t),\quad g_h^\circ(\mathbf{x},t)=\sum_{k=1}^{N} g_k^{h\circ}(t)p_k^h(\mathbf{x}),\\
&g_k^{h\circ}(t) = \int_\Omega \frac{g(\boldsymbol{\xi},t)}{\sqrt{h(\boldsymbol{\xi})}}p_k^h(\boldsymbol{\xi})\,d\Omega_\xi,\quad k=1,\ldots,N.
\end{aligned} \tag{28}$$

Note that in the representations (25)–(28) for $q(\mathbf{x},t)$, the function $q^\circ(\mathbf{x},t)$ is known (as determined by the auxiliary conditions), and the term $q^*(\mathbf{x},t)$ is to be found. Conversely, for the right-hand side, we should find $f_h^\circ(\mathbf{x},t)$, and $f_h^*(\mathbf{x},t)$ is given by $g(\mathbf{x},t)/\sqrt{h(\mathbf{x})}$.

One can introduce an operator of orthogonal projection that maps the space $L_2(\Omega)$ onto $L_2^{h\circ}(\Omega)$:

$$\mathcal{P}_h^\circ f(\mathbf{x}) = \int_\Omega f(\boldsymbol{\xi})\sum_{i=1}^{N} p_i^h(\mathbf{x})p_i^h(\boldsymbol{\xi})\,d\Omega_\xi. \tag{29}$$

Obviously, the orthogonal projector $\mathcal{P}_h^* = \mathcal{I} - \mathcal{P}_h^\circ$ maps $L_2(\Omega)$ onto $L_2^{h*}(\Omega)$. Moreover, the following relations hold:

$$\begin{aligned} &\mathcal{P}_h^\circ q(\mathbf{x},t) = q^\circ(\mathbf{x},t), \quad \mathcal{P}_h^* q(\mathbf{x},t) = q^*(\mathbf{x},t), \\ &\mathcal{P}_h^\circ \frac{f(\mathbf{x},t)}{\sqrt{h(\mathbf{x})}} = f_h^\circ(\mathbf{x},t), \quad \mathcal{P}_h^* \frac{f(\mathbf{x},t)}{\sqrt{h(\mathbf{x})}} = f_h^*(\mathbf{x},t). \end{aligned} \tag{30}$$

We apply the projection operator $\mathcal{P}_h^*$ to equation (20) and obtain an integral equation in $L_2^{h*}(\Omega)$ (with a known right-hand side) for the determination of $q^*(\mathbf{x},t)$:

$$\begin{aligned} &\sigma(t)(\mathcal{I} - \mathcal{V}_1)q^*(\mathbf{x},t) + (\mathcal{I} - \mathcal{V}_2)\mathcal{P}_h^*\mathcal{F}^h q^*(\mathbf{x},t) \\ &\qquad\qquad = -g^*(\mathbf{x},t) - (\mathcal{I} - \mathcal{V}_2)\mathcal{P}_h^*\mathcal{F}^h q^\circ(\mathbf{x},t), \\ &\mathcal{P}_h^*\mathcal{F}^h\phi(\mathbf{x},t) = \int_\Omega F_h^*(\mathbf{x},\boldsymbol{\xi})\phi(\boldsymbol{\xi},t)\,d\Omega_\xi, \quad \mathbf{x}\in\Omega, \quad \tau_0 \le t \le T, \\ &F_h^*(\mathbf{x},\boldsymbol{\xi}) = F^h(\mathbf{x},\boldsymbol{\xi}) - \int_\Omega F^h(\mathbf{s},\boldsymbol{\xi})\sum_{i=1}^N p_i^h(\mathbf{x})p_i^h(\mathbf{s})\,d\Omega_s. \end{aligned} \tag{31}$$

The operator $\mathcal{P}_h^*\mathcal{F}^h$ is a Hilbert–Schmidt operator from $L_2^{h*}(\Omega)$ to $L_2^{h*}(\Omega)$. Let us construct a solution of equation (31) in the form of a series with respect to its eigenfunctions that form a basis in $L_2^{h*}(\Omega)$. Let us construct the system of these functions.

Let $\varphi_k^{h*}(\mathbf{x})$ be eigenfunctions of the operator $\mathcal{P}_h^*\mathcal{F}^h$ and μ_k^{h*} the corresponding eigenvalues. We have

$$\mathcal{P}_h^*\mathcal{F}^h\varphi_k^{h*}(\mathbf{x}) = \mu_k^{h*}\varphi_k^{h*}(\mathbf{x}), \quad k = N+1, N+2, \ldots \tag{32}$$

Let us represent the eigenfunction $\varphi_i^{h*}(\mathbf{x})$ as a series with respect to the basis $p_i^h(\mathbf{x})$ $(i \ge N+1)$

$$\begin{aligned} &\varphi_k^{h*}(\mathbf{x}) = \sum_{i=N+1}^\infty \varphi_{i(k)}^{h*} p_i^h(\mathbf{x}), \quad p_i^h(\mathbf{x}) = \frac{P_i^h(\mathbf{x})}{\sqrt{h(\mathbf{x})}}, \\ &\Phi_k^{h*}(\mathbf{x}) = \sum_{i=N+1}^\infty \varphi_{i(k)}^{h*} P_i^h(\mathbf{x}). \end{aligned} \tag{33}$$

Using (11) and (31), we obtain the following double series expansion for the kernel $F_h^*(\mathbf{x},\boldsymbol{\xi})$:

$$F_h^*(\mathbf{x},\boldsymbol{\xi}) = \sum_{m=N+1}^\infty \sum_{n=N+1}^\infty F_{mn}^h p_m^h(\mathbf{x}) p_n^h(\boldsymbol{\xi}) + \sum_{i=1}^N \sum_{n=N+1}^\infty F_{in}^h p_n^h(\mathbf{x}) p_i^h(\boldsymbol{\xi}). \tag{34}$$

Note that the coefficients in the expansion of the kernel $F_h^*(\mathbf{x}, \boldsymbol{\xi})$ in (34) coincide with those in the expansion of the kernel $F^h(\mathbf{x}, \boldsymbol{\xi})$, and this allows us to use the available data instead of recalculating the coefficients of the new problem.

Substituting (33) and (34) into (32), we obtain an infinite system of linear algebraic equations for the determination of the eigenvalues and the eigenfunction expansion coefficients. This system has a symmetric matrix and can be written in the form

$$\sum_{n=N+1}^{\infty} F_{mn}^h \varphi_{n(k)}^{h*} = \mu_k^{h*} \varphi_{m(k)}^{h*}, \quad m = N+1, N+2, \ldots \tag{35}$$

Now, let us construct a solution of equation (31). To this end, we represent the functions $q^*(\mathbf{x}, t)$ and $g_h^*(\mathbf{x}, t)$ in the form of series with eigenfunctions of the operator $\mathcal{P}_h^* \mathcal{F}^h$:

$$\begin{aligned} q^*(\mathbf{x}, t) &= \sum_{k=N+1}^{\infty} q_k^*(t) \varphi_k^{h*}(\mathbf{x}), \quad g_h^*(\mathbf{x}, t) = \sum_{k=N+1}^{\infty} g_k^{h*}(t) \varphi_k^{h*}(\mathbf{x}), \\ g_k^{h*}(t) &= \int_{-1}^{1} g_h^*(\boldsymbol{\xi}, t) \varphi_k^{h*}(\boldsymbol{\xi}) \, d\Omega_\xi, \end{aligned} \tag{36}$$

and substitute these into (31). Then, taking into account (25)–(28) and (30)–(34), we obtain the following sequence of independent Volterra equations of the second kind:

$$\begin{aligned} &(\mathcal{I} - \mathcal{V}_k^{h*}) q_k^*(t) = \delta_k^{h*}(t), \quad \mathcal{V}_k^{h*} = \frac{\sigma(t)\mathcal{V}_1 + \mu_k^{h*}\mathcal{V}_2}{\sigma(t) + \mu_k^{h*}}, \\ &\mathcal{V}_k^{h*} f(t) = \int_{\tau_0}^{t} V_k^{h*}(t, \tau) f(\tau) \, d\tau, \\ &V_k^{h*}(t, \tau) = \frac{\sigma(t) V_1(t, \tau) + \mu_k^{h*} V_2(t, \tau)}{\sigma(t) + \mu_k^{h*}}, \\ &\delta_k^{h*}(t) = -\frac{1}{\sigma(t) + \mu_k^{h*}} \left[g_k^{h*}(t) + (\mathcal{I} - \mathcal{V}_2) \sum_{i=1}^{N} F_{k(i)}^h q_i^\circ(t) \right], \\ &F_{k(i)}^h = \sum_{n=N+1}^{\infty} F_{in}^h \varphi_{n(k)}^{h*}, \qquad i = 1, \ldots, N, \quad k = N+1, N+2, \ldots \end{aligned} \tag{37}$$

Resolving (37) with respect to $q_k^*(t)$, we get

$$\begin{aligned} &q_k^*(t) = (\mathcal{I} + \mathcal{R}_k^{h*}) \delta_k^{h*}(t), \quad (\mathcal{I} - \mathcal{V}_k^{h*})^{-1} = (\mathcal{I} + \mathcal{R}_k^{h*}), \\ &\mathcal{R}_k^{h*} f(t) = \int_{\tau_0}^{t} R_k^{h*}(t, \tau) f(\tau) \, d\tau, \end{aligned} \tag{38}$$

where $R_k^{h*}(t,\tau)$ is the resolvent of the kernel $V_k^{h*}(t,\tau)$.

We see that in view of (35)–(38) the function $q^*(\mathbf{x},t)$ has been determined, and it is easy to find $q(\mathbf{x},t)$, since $q^\circ(\mathbf{x},t)$ is known by assumption (see (25)–(27)). Hence, taking into account the transformation of the variables (2), we finally obtain

$$y(\mathbf{x},t) = \frac{1}{h(\mathbf{x})}\left[\sum_{n=1}^{N}\sum_{k=1}^{n} a_{nk}M_k(t)P_n^h(\mathbf{x}) + \sum_{k=N+1}^{\infty} q_k^*(t)\Phi_k^{h*}(\mathbf{x})\right]. \tag{39}$$

The solution (39) depends on the function $h(x)$ in explicit manner, and this allows us to solve equation (18) with high accuracy by keeping a relatively small number of terms in the series even for a rapidly oscillating $h(\mathbf{x})$.

Now, in order to find the functions $\alpha_i(t)$ $(i = 1,\ldots,N)$, we apply the projection operator $\mathcal{P}_h^\circ$ to equation (31). As a result, we get

$$\alpha_k(t) = \sum_{i=k}^{N} a_{ik}\left\{ g_i^{h\circ}(t) + \sigma(t)(\mathcal{I}-\mathcal{V}_1)\sum_{m=1}^{i} a_{im}M_m(t) \right.$$
$$\left. +(\mathcal{I}-\mathcal{V}_2)\left[\sum_{j=1}^{N} F_{ji}^h \sum_{m=1}^{j} a_{jm}M_m(t) + \sum_{j=N+1}^{\infty} F_{j(i)}^h q_j^*(t)\right]\right\}. \tag{40}$$

Note that relations (40) form a system of N linear algebraic equations (with a triangular matrix) for the determination of the unknown quantities $\alpha_1(t),\ldots,\alpha_N(t)$.

Thus, we have constructed a complete solution of the integral equation (18) with the auxiliary conditions (19).

3. Application to Contact Mechanics

Let us solve a plane problem of the interaction between a rigid punch and a viscoelastic foundation with rough surface which is described by the rapidly oscillating function $h(x)$. The surface of the punch is described by comparatively smooth function $g(x)$. Viscoelastic behavior of the foundation and the coating are given by continuous function $\sigma(t)$ and continuous or polar kernels $V_1(t,\tau)$ and $V_2(t,\tau)$ respectively.

Consider equation (1) with the right-hand side in the form $f(x,t) = \alpha_1(t)+\alpha_2(t)x-g(x,t)$ and two auxiliary integral conditions on the unknown function $y(x,t)$. The problem is to find the solution of the integral equation

$(-1 \le x \le 1,\ 1 \le t \le T)$

$$\sigma(t)\left[y(x,t) - \int_1^t V_1(t,\tau) y(x,\tau)\,d\tau\right] + \int_{-1}^{1} \frac{F(x,\xi)}{h(x)} y(\xi,t)\,d\xi$$
$$- \int_1^t V_2(t,\tau) \int_{-1}^{1} S(x,\xi) y(\xi,\tau)\,d\xi\,d\tau = \frac{\alpha_1(t)}{h(x)} + \frac{\alpha_2(t)x}{h(x)} - \frac{g(x,t)}{h(x)}, \tag{41}$$

with the auxiliary conditions

$$\int_{-1}^{1} y(\xi,t)\,d\xi = M_1(t), \qquad \int_{-1}^{1} \xi\, y(\xi,t)\,d\xi = M_2(t), \tag{42}$$

where $y(x,t)$ is the contact pressure, $\alpha_1(t)$ and $\alpha_2(t)$ are the unknown settlement and tilt angle of a punch, $F(x,\xi)$ is the known kernel of the plane contact problem compact operator, x is the longitudinal coordinate, and t is the time variable.

Let us transform the equation with the Schmidt kernel to an equation with a Hilbert–Schmidt kernel by changing the variables in (41) and (42) (see (2)). As a result, the integral equation and the auxiliary conditions become $(-1 \le x \le 1,\ 1 \le t \le T)$

$$\sigma(t)\left[q(x,t) - \int_1^t V_1(t,\tau) q(x,\tau)\,d\tau\right] + \int_{-1}^{1} F^h(x,\xi) q(\xi,t)\,d\xi \tag{43}$$
$$- \int_{\tau_0}^t V_2(t,\tau) \int_{-1}^{1} F^h(x,\xi) q(\xi,\tau)\,d\xi\,d\tau = \frac{\alpha_1(t)}{\sqrt{h(x)}} + \frac{\alpha_2(t)x}{\sqrt{h(x)}} - \frac{g(x,t)}{\sqrt{h(x)}},$$

$$\int_{-1}^{1} \frac{q(\xi,t)}{\sqrt{h(\xi)}}\,d\xi = M_1(t), \qquad \int_{-1}^{1} \frac{q(\xi,t)}{\sqrt{h(\xi)}}\xi\,d\xi = M_2(t). \tag{44}$$

In order to construct a solution of the mixed integral equation (43) with the auxiliary conditions (44), we use the basis $p_n^h(x)$ in $L_2[-1,1]$ and note that the space $L_2[-1,1]$ can be represented as a direct sum of its orthogonal subspaces, $L_2[-1,1] = L_2^{h\circ}[-1,1] \oplus L_2^{h*}[-1,1]$, where $L_2^{h\circ}[-1,1]$ is the Euclidean space with the basis $p_1^h(x)$ and $p_2^h(x)$, and $L_2^{h*}[-1,1]$ is the Hilbert space with the basis $p_k^h(x)$ $(k = 3, 4, \ldots)$. It can be seen that the integrand and the right-hand side can be represented as a sum of continuous functions of $t \in [1,T]$ with values in $L_2^{h\circ}[-1,1]$ and $L_2^{h*}[-1,1]$, respectively, i.e,

$$q(x,t) = q^\circ(x,t) + q^*(x,t), \qquad \frac{f(x,t)}{\sqrt{h(x)}} = f_h^\circ(x,t) + f_h^*(x,t). \tag{45}$$

Note that in the representation (45) for $q(x,t)$, the function $q^\circ(x,t)$ is known as determined by the auxiliary conditions, and the term $q^*(x,t)$

should be found. Conversely, for the right-hand side, $f_h^\circ(x,t)$ should be found and $f_h^*(x,t)$ is determined by the function $g(x,t)/\sqrt{h(x)}$. The facts mentioned above allow us to classify the resulting problem as a special case of the general projection problem.

According to the general method, in the present case one can introduce an operator of orthogonal projection that maps the space $L_2[-1,1]$ onto $L_2^{h\circ}[-1,1]$:

$$\mathcal{P}_h^\circ \phi(x,t) = \int_{-1}^{1} \phi(\xi,t)[p_1^h(x)p_1^h(\xi) + p_2^h(x)p_2^h(\xi)]\, d\xi. \tag{46}$$

Obviously, the projector $\mathcal{P}_h^* = \mathcal{I} - \mathcal{P}_h^\circ$ maps $L_2[-1,1]$ onto $L_2^{h*}[-1,1]$. Moreover, the following relations hold:

$$\begin{aligned} &\mathcal{P}_h^\circ q(x,t) = q^\circ(x,t), \quad \mathcal{P}_h^* q(x,t) = q^*(x,t), \\ &\mathcal{P}_h^\circ \frac{f(x,t)}{\sqrt{h(x)}} = f_h^\circ(x,t), \quad \mathcal{P}_h^* \frac{f(x,t)}{\sqrt{h(x)}} = f_h^*(x,t), \end{aligned} \tag{47}$$

Let us apply the projection operator $\mathcal{P}_h^*$ to equation (43). Then, for the determination of $q^*(x,t)$, we obtain the integral equation in $L_2^{h*}[-1,1]$ with a known right-hand side and the Hilbert–Schmidt kernel in the form

$$F_h^*(x,\xi) = F^h(x,\xi) - \int_{-1}^{1} F^h(s,\xi)[p_1^h(x)p_1^h(s) + p_2^h(x)p_2^h(s)]\, ds \tag{48}$$

The solution of this equation can be constructed in the form of a series in terms of eigenfunctions of the kernel (48) which form a basis in the Hilbert space $L_2^{h*}[-1,1]$. For an eigenfunction $\varphi_k^{h*}(x)$, let μ_k^{h*} be the corresponding eigenvalue of the kernel $F_p^*(x,\xi)$. Then

$$\int_{-1}^{1} F_h^*(x,\xi)\varphi_k^{h*}(\xi)\, d\xi = \mu_k^{h*}\varphi_k^{h*}(x), \quad k = 3,4,\ldots \tag{49}$$

Now, let us construct a solution of obtained equation using (49). To that end, we represent the functions $q^*(x,t)$ and $g_h^*(x,t)$ as series in terms of eigenfunctions of the kernel $F_h^*(x,\xi)$:

$$\begin{aligned} &q^*(x,t) = \sum_{k=3}^{\infty} q_k^*(t)\varphi_k^{h*}(x), \quad g_h^*(x,t) = \sum_{k=3}^{\infty} g_k^{h*}(t)\varphi_k^{h*}(x), \\ &g_k^{h*}(t) = \int_{-1}^{1} g_h^*(x,t)\varphi_k^{h*}(x)\, dx. \end{aligned} \tag{50}$$

Substituting (50) into the integral equation, we obtain the following sequence of independent Volterra equations of the second kind:

$$q_k^*(t) - \int_1^t V_k^{h*}(t,\tau) q_k^*(\tau)\, d\tau = \delta_k^{h*}(t), \quad i = 1, 2, \quad k = 3, 4, \ldots,$$
$$V_k^{h*}(t,\tau) = \frac{\sigma(t)V_1(t,\tau) + \mu_k^{h*} V_2(t,\tau)}{\sigma(t) + \mu_k^{h*}},$$
$$\delta_k^{h*}(t) = -\frac{1}{\sigma(t) + \mu_k^{h*}} \left[g_k^{h*}(t) + \sum_{i=1}^{2} F_{k(i)}^h q_i^\circ(t) \right. \tag{51}$$
$$\left. - \int_1^t V_2(t,\tau) \sum_{i=1}^{2} F_{k(i)}^h q_i^\circ(\tau)\, d\tau \right], \quad F_{k(i)}^h = \sum_{n=3}^{\infty} F_{in}^h \varphi_{n(k)}^{h*},$$

Resolving (51) with respect to $q_k^*(t)$ by any known method, we obtain

$$q_k^*(t) = f_k^{h*}(t) + \int_1^t R_k^{h*}(t,\tau) f_k^{h*}(\tau)\, d\tau, \tag{52}$$

where $R_k^{h*}(t,\tau)$ is the resolvent of the kernel $V_k^{h*}(t,\tau)$.

Now, in view of (52), the function $q^*(x,t)$ has been determined, and therefore, we easily find $q(x,t)$, since $q^\circ(x,t)$ is known.

Now, let us find the functions $\alpha_1(t)$ and $\alpha_2(t)$. To this end, we apply the projection operator $\mathcal{P}_h^\circ$ to the integral equation. As a result, we obtain the system of two linear algebraic equations (with a triangular matrix) for the determination of the unknown quantities $\alpha_1(t)$ and $\alpha_2(t)$.

4. Conclusions

Thus, we have constructed the solution of integral equation with two Volterra operators and one compact operator. New proposed projection method has allowed us

- obtain such a form of the solution where function h enters the solution in explicit form,
- avoid the solution of infinite system of integral Volterra equations to which the classical method of separation of variables leads,
- solve mixed integral equation with high accuracy even for a rapidly oscillating function h when the classical method gives up to 100% mistake.

Acknowledgments

This work was supported by the Russian Foundation for Basic Research under the Project RFBR–BRICS.

References

1. L. A. Galin and G. M. L. Gladwell (Editor), *Contact Problems: The legacy of L. A. Galin* (Springer, Dordrecht, 2008).
2. I. G. Goryacheva, *Contact Mechanics in Tribology* (Springer, Dordrecht, 1998).
3. A. V. Manzhirov, Axisymmetric contact problems for non-uniformly aging layered viscoelastic foundations, *J Appl Math Mech* **47**, 4 (1983)
4. A. V. Manzhirov, On a method of solving two-dimensional integral equations of axisymmetric contact problems for bodies with complex rheology, *J Appl Math Mech* **49**, 6 (1985)
5. V. M. Alexandrov and A. V. Manzhirov, Two-dimensional integral equations in applied mechanics of solids, *J Appl Mech Tech Phys* **28**, 5 (1987)
6. A. V. Manzhirov and V. A. Chernysh, On the interaction of a rigid reinforcing sleeve and inhomogeneous aging high-pressure pipes, *Mech Solids* **22**, 6 (1988)
7. A. V. Manzhirov and V. A. Chernysh, Contact problem for a layered inhomogeneous aging cylinder reinforced by a rigid ring, *J Appl Mech Tech Phys* **30**, 6 (1990)
8. N. Kh. Arutyunyan, A. V. Manzhirov, and V. E. Naumov *Contact Problems in Mechanics of Growing Bodies [in Russian]* (Nauka Publ., Moscow, 1991).
9. N. Kh. Arutyunyan and A. V. Manzhirov, *Contact Problems in the Theory of Creep [in Russian]* (Izd-vo Inst. Mekaniki NAN Armenii, Yerevan, 1999).
10. K. E. Kazakov and A. V. Manzhirov, Conformal contact between layered foundations and punches, *Mech Solids* **32**, 3 (2008)
11. A .V. Manzhirov, A mixed integral equation of mechanics and a generalized projection method of its solution, *Doklady Physics* **61**, 10 (2016)
12. A .V. Manzhirov, Integral equations with several different operators and their application to mechanics, *Lecture Notes in Engineering and Computer Science: Proceedings of The World Congress on Engineering 2016, WCE 2016, 29 June-1 July, 2016, London, U.K., pp. 10–15.*

SOLVING FOR ANALYTICAL SOLUTIONS OF THE NAVIER-STOKES MODEL EQUATIONS USING HE'S POLYNOMIALS APPROACH*

S. O. EDEKI†, *Member, IAENG* and G. O. AKINLABI‡
Department of Mathematics, Covenant University,
†soedeki@yahoo.com
‡grace.akinlabi@covenantuniversity.edu.ng

M. E. ADEOSUN
Department of Mathematics and Statistics, Osun State College of Technology, Esa-Oke, Nigeria
mabelruore@gmail.com

Navier-Stokes models are of great importance in applied sciences. In this paper, He's polynomials approach is implemented for obtaining approximate and exact solutions of the Navier-Stokes model equations. These solutions are calculated in the form of series with easily computable components. This technique is showed to be very effective, efficient and reliable because it gives the exact solution of the solved problems with less computational work, without neglecting the level of accuracy. We therefore, recommend the extension and application of this novel method for solving problems arising in other aspect of applied sciences. Numerical computations, and graphics done in this work, are through Maple 18.

Keywords: Analytical solutions; He's polynomials; Navier-Stokes model.

1. Introduction

Navier-Stokes equations are fundamental models in applied sciences such as physics. They are used in describing the motion of viscous fluid substances. These models are very useful as they describe the physics of many phenomena relating mathematics, engineering, pure and applied sciences. In computational fluid dynamics, Navier-Stokes equations are the main equations, relating pressure and external forces acting on fluid to the response of the fluid flow [1].

In general form, the Navier-Stokes and continuity equations are given by:

* Work partially supported by CUCRID: Covenant University Centre for Research, Innovation and Discovery.

$$\frac{\partial u}{\partial t}+(\underline{u}\cdot\nabla)\underline{u}=-\frac{1}{\rho}\nabla P+v\nabla^2\underline{u} \tag{1.1}$$

$$\nabla\cdot\underline{u}=0 \tag{1.2}$$

where u is the flow velocity, $\underline{u}$ is the velocity, v is the kinematics viscosity, P is the pressure, t is the time, ρ is the density, and ∇ is a del operator.

In considering unsteady, one dimensional motion of a viscous fluid in a tube; the equations of motion governing the flow field in the tube are Navier-Stokes equations in cylindrical coordinates [1]. These are denoted by:

$$\frac{\partial u}{\partial t}=P+v\left(\frac{\partial^2 u}{\partial r^2}+\frac{1}{r}\frac{\partial u}{\partial r}\right) \tag{1.3}$$

subject to:

$$u(r,0)=g(r) \tag{1.4}$$

where $P=-\frac{\partial P}{\rho\partial z}$.

In a bid to providing numerical and/or exact solutions to linear and nonlinear differential equations; many researchers have considered and developed a lot of semi-analytical methods. These include Adomian Decomposition Method (ADM), Variational Iteration Method (VIM), Differential Transform Method (DTM), Laplace Transform Method (LTM), and their modified forms [3-7].

Recently, He [8] developed the Homotopy Perturbation Method (HPM) for solving differential equations. Basically, the merit of the HPM is to overcome the difficulties involved in calculating the nonlinear terms in the concerned problem. The HPM has wider applications when handling different classes of differential equations, integral equations, integro-differential equations and so on [9-15]. As a modification of the HPM, Ghorbani *et al.* [16,17] introduced the He's polynomials where nonlinear terms were split into series of polynomials which are calculated from HPM. It is remarked that He's polynomials are attuned with Adomian's polynomials, yet it is showed that the He's polynomials are easier to compute, and are very much user friendly.

In considering the solutions of the Navier-Stokes equation (of integer, and non-integer order), some of the semi-analytical methods have been applied [1-3, 18-21].

In this work, it is therefore, our intention to extend the conference paper of Edeki *et al.* [22] by providing analytical solutions to the Navier–Stokes model as contained in (1.1)-(1.3) via the means of He's polynomials method . It is worth mentioning that the He's polynomials method is an alternative semi-analytical method, even without giving up accuracy.

2. The overview of the He's Polynomial Method [16, 17, 22-26]

Suppose $\Im$ is an integral or a differential operator, then we consider the equation of the form:

$$\Im(\varpi)=0 \tag{2.1}$$

Let $H(\varpi,p)$ be a convex homotopy defined by:

$$H(\varpi,p)=p\Im(\varpi)+(1-p)G(\varpi) \tag{2.2}$$

where $G(\varpi)$ is a functional operator with ϖ_0 as a known solution. Thus, we have:

$$H(\varpi,0)=G(\varpi) \text{ and } H(\varpi,1)=\Im(\varpi) \tag{2.3}$$

whenever $H(\varpi,p)=0$ is satisfied, and $p\in(0,1]$ is an embedded parameter. In HPM, p is used as an expanding parameter to obtain:

$$\varpi=\sum_{j=0}^{\infty}p^j\varpi_j=\varpi_0+p\varpi_1+p^2\varpi_2+\cdots \tag{2.4}$$

From (2.4), the solution is obtained as $p\to 1$. The convergence of (2.4) as $p\to 1$ has been considered in [23].

The method considers $N(\varpi)$ (the nonlinear term) as:

$$N(\varpi)=\sum_{j=0}^{\infty}p^jH_j=H_o+p^1H_1+p^2H_2+\cdots \tag{2.5}$$

where H_k's are the so-called He's polynomials, which can be computed using:

$$H_k(\varpi)=\frac{1}{k!}\frac{\partial^k}{\partial p^k}\left(N\left(\sum_{j=0}^{k}p^j\varpi_j\right)\right)_{p=0},\ n\geq 0 \tag{2.6}$$

where $H_k(\varpi)=H_k(\varpi_0,\varpi_1,\varpi_2,\varpi_3,\ldots,\varpi_k)$.

3. The He's Polynomials and the Navier-Stokes model

In this subsection, the He's Polynomials approach will be applied to the following Navier-Stokes model as follows:

A Problem 1: Consider the following Navier-Stokes model:

$$\frac{\partial w}{\partial t}=\frac{\partial^2 w}{\partial r^2}+\frac{1}{r}\frac{\partial w}{\partial r} \tag{3.1}$$

subject to:

$$w(r,0)=r \tag{3.2}$$

Procedure w.r.t Problem 1:

We re-write (3.1) in an integral form as in (3.3) below:

$$w(r,t)=w(r,0)+\int_0^t\left\{\frac{\partial^2 w}{\partial r^2}+\frac{1}{r}\frac{\partial w}{\partial r}\right\}dt \tag{3.3}$$

Applying the convex homotopy method to (3.3) gives:

$$\sum_{n=0}^{\infty}p^n w_n=w(r,0)+p\int_0^t\left\{\sum_{n=0}^{\infty}p^n\frac{\partial^2 w_n}{\partial r^2}+\frac{1}{r}\sum_{n=0}^{\infty}p^n\frac{\partial w_n}{\partial r}\right\}dt \tag{3.4}$$

Further simplification of (3.4) gives:

$$w_0+pw_1+p^2w_2+\cdots=r+p\int_0^t\left\{\left(\frac{\partial^2 w_0}{\partial r^2}+p\frac{\partial^2 w_1}{\partial r^2}+p^2\frac{\partial^2 w_2}{\partial r^2}+\cdots\right)\right.$$
$$\left.+\frac{1}{r}\left(\frac{\partial w_0}{\partial r}+p\frac{\partial w_1}{\partial r}+p^2\frac{\partial w_2}{\partial r}+\cdots\right)\right\}dt \tag{3.5}$$

From (3.5), comparing the coefficients of the equal powers of p in the following ways gives:

$$p^{(0)}:w_0=r$$

$$p^{(1)}:w_1=\int_0^t\left(\frac{\partial^2 w_0}{\partial r^2}+\frac{1}{r}\frac{\partial w_0}{\partial r}\right)dt$$

$$p^{(2)}:w_2=\int_0^t\left(\frac{\partial^2 w_1}{\partial r^2}+\frac{1}{r}\frac{\partial w_1}{\partial r}\right)dt$$

$$p^{(3)}:w_3=\int_0^t\left(\frac{\partial^2 w_2}{\partial r^2}+\frac{1}{r}\frac{\partial w_2}{\partial r}\right)dt$$

$$\vdots$$

$$p^{(k)}:w_k=\int_0^t\left(\frac{\partial^2 w_{k-1}}{\partial r^2}+\frac{1}{r}\frac{\partial w_{k-1}}{\partial r}\right)dt,\ k\geq 1$$

Therefore, to obtain $w_1, w_2, w_3, \ldots$, we will apply $w_0 = r$ for the simplification of $p^{(1)}$, $p^{(2)}$, $p^{(3)}$, and so on, thus:

$$w_0 = r,\ w_1 = \frac{t}{r},\ w_2 = \frac{1}{2}\frac{t^2}{r^3},\ w_3 = \frac{3}{2}\frac{t^3}{r^5},\ w_4 = \frac{75}{8}\frac{t^4}{r^7},$$

$$w_5 = \frac{735}{8}\frac{t^5}{r^9},\ w_6 = \frac{19845}{16}\frac{t^6}{r^{11}},\ w_7 = \frac{343035}{16}\frac{t^7}{r^{13}}\cdots$$

$$\therefore\ w(x,t) = w_0 + w_1 + w_2 + w_3 + w_4 + w_5 + \cdots$$

$$= r + \frac{t}{r} + \frac{1}{2}\frac{t^2}{r^3} + \frac{3}{2}\frac{t^3}{r^5} + \frac{75}{8}\frac{t^4}{r^7} + \frac{735}{8}\frac{t^5}{r^9}$$

$$+ \frac{19845}{16}\frac{t^6}{r^{11}} + \frac{343035}{16}\frac{t^7}{r^{13}} + \cdots$$

$$= r + \sum_{j=1}^{\infty} \frac{1^1 \times 3^2 \times 5^2 \times \cdots \times (2j-3)^2}{r^{2j-1}} \frac{t^j}{j!} \tag{3.6}$$

Remark: For graphical consideration of the approximate solution, we use $r \in [0, 0.5]$ and $t \in [0, 0.2]$. Figures 1 and 2 represent the 3D plots of the solution (*of problem* 1) for terms up to power seven and power five (in terms of the time variable t) respectively.

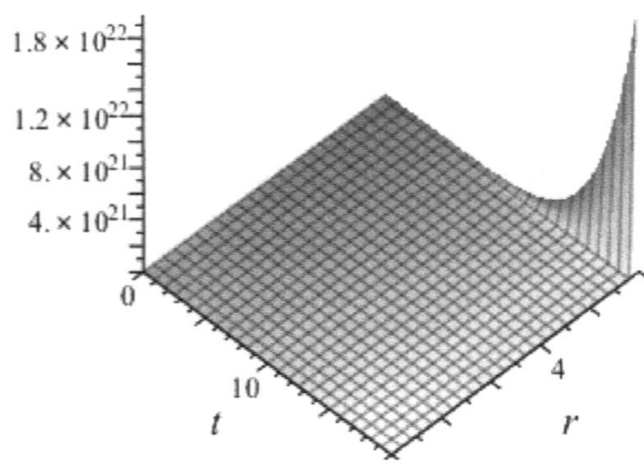

Fig. 1. He's polynomial solution up to t^7.

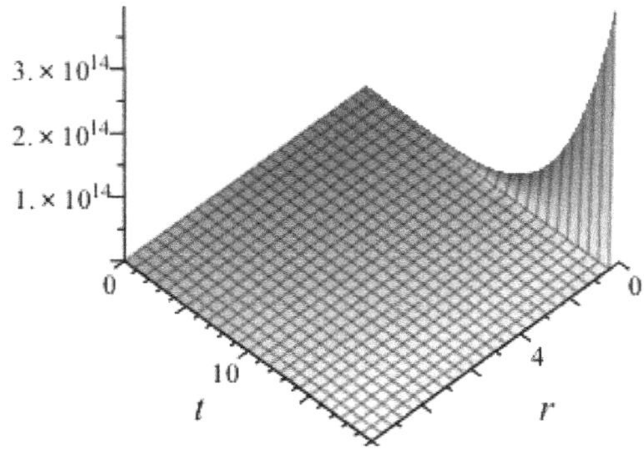

Fig. 2. He's polynomial solution up to t^5.

B. Problem 2: Consider the following Navier-Stokes model:

$$\frac{\partial w}{\partial t} = p + \frac{\partial^2 w}{\partial r^2} + \frac{1}{r}\frac{\partial w}{\partial r} \tag{3.7}$$

subject to:

$$w(r,0) = 1 - r^2 \tag{3.8}$$

Procedure w.r.t Problem 2:

We re-write (3.7) in an integral form as in (3.9) below:

$$w(r,t) = w(r,0) + \int_0^t \left\{ p + \frac{\partial^2 w}{\partial r^2} + \frac{1}{r}\frac{\partial w}{\partial r} \right\} dt \tag{3.9}$$

Applying the convex homotopy method to (3.9) gives:

$$\sum_{n=0}^{\infty} h^n w_n = w(r,0) + h\int_0^t \left\{ p + \sum_{n=0}^{\infty} h^n \frac{\partial^2 w_n}{\partial r^2} + \frac{1}{r}\sum_{n=0}^{\infty} h^n \frac{\partial w_n}{\partial r} \right\} dt \tag{3.10}$$

Further simplification of (3.10) gives:

$$w_0 + hw_1 + h^2 w_2 + \cdots = \left(1 - r^2\right) + h\int_0^t \left\{ p + \left(\frac{\partial^2 w_0}{\partial r^2} + h\frac{\partial^2 w_1}{\partial r^2} + h^2 \frac{\partial^2 w_2}{\partial r^2} + \cdots \right) \right.$$

$$+\frac{1}{r}\left(\frac{\partial w_0}{\partial r}+h\frac{\partial w_1}{\partial r}+h^2\frac{\partial w_2}{\partial r}+\cdots\right)\Bigg\}\,dt \tag{3.11}$$

From (3.11), comparing the coefficients of the equal powers of h in the following ways gives:

$$h^{(0)}: w_0 = 1-r^2$$

$$h^{(1)}: w_1 = \int_0^t \left(p+\frac{\partial^2 w_0}{\partial r^2}+\frac{1}{r}\frac{\partial w_0}{\partial r}\right)dt$$

$$h^{(2)}: w_2 = \int_0^t \left(\frac{\partial^2 w_1}{\partial r^2}+\frac{1}{r}\frac{\partial w_1}{\partial r}\right)dt$$

$$h^{(3)}: w_3 = \int_0^t \left(\frac{\partial^2 w_2}{\partial r^2}+\frac{1}{r}\frac{\partial w_2}{\partial r}\right)dt$$

$$\vdots$$

$$h^{(k)}: w_k = \int_0^t \left(\frac{\partial^2 w_{k-1}}{\partial r^2}+\frac{1}{r}\frac{\partial w_{k-1}}{\partial r}\right)dt,\ k\geq 2$$

Therefore, to obtain $w_1, w_2, w_3, \ldots$, we will apply $w_0 = 1-r^2$ for the simplification of $p^{(1)}$, $p^{(2)}$, $p^{(3)}$, and so on, thus:

$$w_0 = 1-r^2, w_1 = (p-4)t, w_\eta = 0 \text{ for } \eta \geq 2$$

$$\therefore\ w(x,t) = w_0 + w_1 + w_2 + w_3 + w_4 + w_5 + \cdots$$

$$= (1-r^2)+(p-4)t \tag{3.12}$$

Remark: for $p=1$, the solution is:

$$w(r,t) = -r^2 - 3t + 1 \tag{3.13}$$

For graphical consideration of the obtained solution, we use $r \in [-1, 1.5]$ and $t \in [0,1]$. Figures 3 and 4 represent *3D plot* & *implicit plot* (respectively) of the solution (*of problem* 2) at $p=1$.

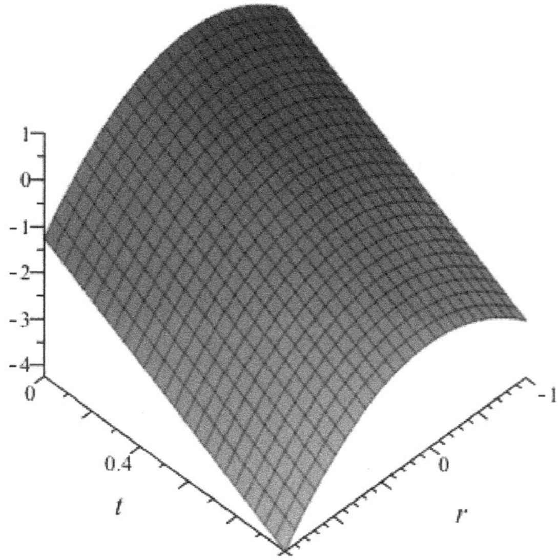

Fig. 3. He's polynomial solution (*3D plot*).

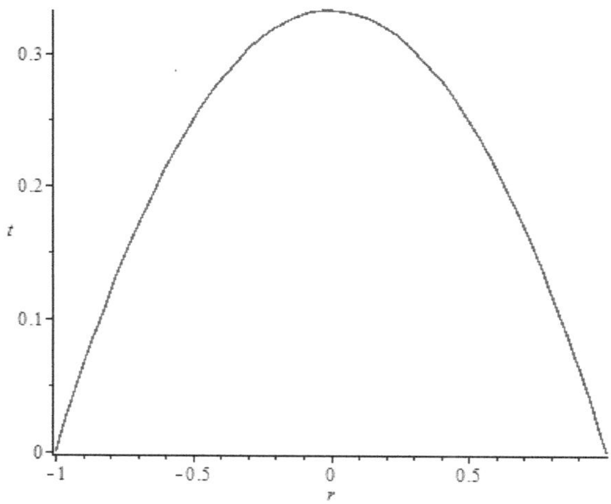

Fig. 4. He's polynomial solution (*implicit plot*).

4. Concluding Remarks

In this paper, a proposed computational method known as He's Polynomials Technique (HPT) has been applied successfully to the Navier-Stokes model for approximate and exact solutions. These solutions were calculated in the form of series with easily computable components. This technique is very much effective, efficient and reliable as it gives the exact solution of the solved problems with less computational work without neglecting the level of accuracy. As a way of comparison, our results are in very much in agreement with those obtained in [1, 2, 21] for a time-order one. We therefore, recommend an extension and application of this novel method for solving problems arising in other areas of applied sciences. Numerical computations, and graphics done in this work, are through Maple 18.

Acknowledgments

The authors wish to thank Covenant University for financial support and provision of good working environment. All constructive and helpful remarks from the anonymous referee(s) are highly appreciated.

References

1. D. Kumar, J. Singh and S Kumar, "A Fractional Model of Navier–Stokes Equation Arising in Unsteady Flow of a Viscous Fluid", *Journal of the Association of Arab Universities for Basic and Applied Sciences* **17**, (2015): 14–19.
2. S. Kumar, D. Kumar, S. Abbasbandy and M.M. Rashidi, "Analytical Solution of Fractional Navier–Stokes Equation by Using Modified Laplace Decomposition Method", *Ain Shams Engineering Journal* **5**, (2014): 569–574.
3. A. M. Wazwaz and M.S. Mehanna, 'The Combined Laplace–Adomian Method for Handling Singular Integral Equation of Heat Transfer", *Int J Nonlinear Sci.* **10** (2010): 248–52.
4. Y. Khan and N. Faraz, "Application of Modified Laplace Decomposition Method for Solving Boundary Layer Equation", *J King Saud Univ,* **23**, (2011):115–9.
5. N. Faraz, "Study of the Effects of the Reynolds Number on Circular Porous Slider via Variational Iteration Algorithm-II", *Computers & Mathematics with Applications* **61**(8), (2011).
6. M. M. Rashidi, "The Modified Differential Transform Method for solving MHD Boundary-layer Equations', *Computer Physics Communications,* **180**, (2009): 2210–7.

7. S. O. Edeki, O. O. Ugbebor and E. A. Owoloko, "Analytical Solutions of the Black–Scholes Pricing Model for European Option Valuation via a Projected Differential Transformation Method", *Entropy,* **17** (11), (2015): 7510–7521.
8. J. H. He, "Homotopy perturbation technique", *Computer Methods in Applied Mechanics and Engineering*, **178** (3/4), (1999): 257–262.
9. J. H. He, "A Coupling Method of Homotopy Techniques and Perturbation Technique for Nonlinear Problems", *International Journal of Non-Linear Mechanics*, **35**(1), (2000): 37–43.
10. M. Shakil, T. Khan, H. A. Wahab and S. Bhatti, "A Comparison of Adomian Decomposition Method (ADM) and Homotopy Perturbation Method (HPM) for Nonlinear Problems", *International Journal of Research in Applied, Natural and Social Sciences*, **1**(3), (2013): 37–48.
11. S. O. Edeki, I. Adinya and O. O. Ugbebor, "The Effect of Stochastic Capital Reserve on Actuarial Risk Analysis via an Integro-differential equation", *IAENG International Journal of Applied Mathematics,* **44** (2), (2014): 83–90.
12. J. Singh, D. Kumar and S. Rathore, "Application of Homotopy Perturbation Transform Method for Solving Linear and Nonlinear Klein-Gordon Equations", *Journal of Information and Computing Science,* **7** (2), (2012): 131–139.
13. N. H. Sweilam and M. M. Khader, "Exact Solutions of some Coupled Nonlinear Partial Differential Equations Using the Homotopy Perturbation Method', *Computers & Mathematics with Applications*, **58**, (2009): 2134–2141.
14. J. Saberi-Nadjafi and A. Ghorbani, "He's Homotopy Perturbation Method: an Effective Tool for Solving Nonlinear Integral and Integro-differential Equations", *Computers & Mathematics with Applications*, **58**, (2009):1345–1351.
15. S. O. Edeki, G. O. Akinlabi and S. A. Adeosun, "On a Modified Transformation Method for Exact and Approximate Solutions of Linear Schrödinger Equations", *AIP Conference Proceedings* **1705**, 020048 (2016); doi: 10.1063/1.4940296.
16. A. Ghorbani and J. S. Nadjfi, "He's Homotopy Perturbation Method for Calculating Adomian's Polynomials", *International Journal of Nonlinear Sciences and Numerical Simulation* **8** (2) (2007): 229–332.
17. A. Ghorbani, "Beyond Adomian's Polynomials: He Polynomials", *Chaos, Solitons & Fractals,* (2007), in press.
18. H. A. Wahab, A. Jamal, S. Bhatti, M. Naeem, M. Shahzad and S. Hussain, "Application of Homotopy Perturbation Method to the Navier-Stokes

Equations in Cylindrical Coordinates", *Computational Ecology and Software*, **5** (2), (2015): 139–151.

19. K. Haldar, "Application of Adomian Approximations to the Navier-Stokes Equation in Cylindrical Coordinates", *Applied Mathematics Letters,* **9** (4), (1995):109–113.
20. N.A Khan, A. Ara, S.A. Ali and A. Mahmood, "Analytical Study of Navier–Stokes Equation with Fractional Orders using He's Homotopy Perturbation and Variational Iteration Methods" *International Journal of Nonlinear Sciences and Numerical Simulation,* **10** (9), (2011):1127–34.
21. S. Momani and Z. Odibat, "Analytical Solution of a Time-Fractional Navier–Stokes Equation by Adomian Decomposition Method", *Applied Mathematics and Computation*, **177**, (2006): 488–94.
22. S. O. Edeki, G. O. Akinlabi and M. E. Adeosun, "Analytical Solutions of the Navier-Stokes Model by He's Polynomials, " Lecture Notes in Engineering and Computer Science: *Proceedings of the World Congress on Engineering* 2016, WCE 2016, June 29-July 1, 2016, London, U.K., pp.16–19.
23. S. O. Edeki, O. O. Ugbebor and E. A. Owoloko, "He's Polynomials for Analytical Solutions of the Black-Scholes Pricing Model for Stock Option Valuation," Lecture Notes in Engineering and Computer Science: *Proceedings of the World Congress on Engineering* 2016, WCE 2016, June 29-July 1, 2016, London, U.K., pp. 632–634.
24. J. H. He, "Homotopy perturbation method: A New Nonlinear Analytical Technique", *Appl. Math. Comput.* **135** (2003):73–79.
25. A. El-Ajou, Z. Odibat , S. Momani and A. Alawneh, "Construction of Analytical Solutions to Fractional Differential Equations Using Homotopy Analysis Method" IAENG *International Journal of Applied Mathematics*, **40** (2), (2010): 43–51.
26. A. Cheniguel and M. Reghioua, "Homotopy Perturbation Method for Solving Some Initial Boundary Value Problems with Non Local Conditions," Lecture Notes in Engineering and Computer Science: *Proceedings of the World Congress on Engineering and Computer Science* 2013, WCECS 2013, 23-25 October, 2013, San Francisco, USA., pp. 572–577.

A GENERAL PURPOSE PROBABILISTIC INFERENCE ENGINE

FENG-JEN YANG
Computer Science and Information Technology Department,
Florida Polytechnic University,
Lakeland, Florida 33805, USA
fyang@flpoly.org
www.floridapolytechnic.org

Our daily life is full of uncertainty. Much of the time, we are thinking, reasoning and making decisions based on the processing of uncertain information. As a result, having a software tool that is able to quantify the likelihood of every hypothesis that we are considering could be really helpful in terms of making informed decisions. In this article, the design and implementation of a domain independent probabilistic inference engine is presented. This inference engine is able to compute the probability that a hypothesis is true based on the actual evidences that are being observed. A demonstration of applying this inference engine to perform probabilistic inferences is also presented. This inference engine has been well tested and ready to be plugged into future systems to speed up their development life cycles.

Keywords: Probabilistic Inference engine, Bayesian inference; Bayesian theory.

1. Introduction

Uncertainty and confusion always go together like the two sides of the same coin. It is inevitable that we all have to cope ourselves with the capability of making smart decisions to reduce the impact of unexpected outcomes. Typically examples, such as medical diagnosis and treatment, stocking market investment, weather forecast, risk management, and so on, are full of our daily life. While we are puzzled by the unsure future outcomes but still have to invest our time and effort to keep going, the need of a scientific approach to compute the probability of every possible outcome becomes very essential. This study is not aiming at inventing a new theory in probability or statistics, but emphasizing on implementing a software tool to apply Bayesian theory as a probabilistic inference engine.

Bayesian theory, a posthumous theory of Thomas Bayes, has been commonly cited in most of the artificial intelligence and machine learning literatures. This is a well known mathematical rationale that supports the drawing of probabilistic conclusions based on the evidences that have been observed.[2] This kind of computation is typically represent by the the following conditional probability:

$$P(H|E) = \frac{P(H \cap E)}{P(E)}$$

In the context of probabilistic inference, this conditional probability is usually interpreted as:

If the evidence E is shown to be true, how much likely is the hypothesis H true?

While designing a system, if a domain expert can provide the values of $P(H \cap E)$ and $P(E)$ then the inference of $P(H|E)$ is just a simple division. However, $P(H \cap E)$ and $P(E)$ are not simple values that human experts can easily construct mentally. As a result, most of the knowledge engineers are trying an indirect knowledge extraction to compute $P(H|E)$ as follows:

$$\begin{aligned} P(H|E) &= \frac{P(H \cap E)}{P(E)} \\ &= \frac{P(E|H) \times P(H)}{P(E|H) \times P(H) + P(E|\bar{H}) \times P(\bar{H})} \end{aligned}$$

Where $P(H)$ is the prior probability of H being true, $P(\bar{H})$ is the prior probability of H being false, $P(E|H)$ is the conditional probability of E being true when H is given to be true, $P(E|\bar{H})$ is the conditional probability of E being true when H is given to be false, and $P(H|E)$ is the posterior probability of H being true when E is seen to be true.

The above computation explains the very fundamental concept of applying Bayesian theory as a probabilistic inference method. However, this is rarely the case that we will be dealing with in real life. In most of the real life applications, we will have to take multiple evidences and multiple hypotheses into consideration. To this end, a generalized computation to accommodate real life applications is presented in the next section.

2. The Generalized Probabilistic Reasoning Method

By taking multiple evidences and multiple hypotheses into consideration, a generalized probabilistic inference steps can be represented as a series of mathematical derivations.

Considering a given problem domain with n evidences and m hypotheses and the inference of:

If evidences E_1, E_2, ... *and* E_n *are shown to be true, how much likely is the hypothesis* Hi *true, where* $1 \leq i \leq m$*?*

The detail inference steps can be performed by the following computations:[1]

$$\begin{aligned}
&P(H_i|(E_1 \cap E_2 \cap \cdots \cap E_n)) \\
&= \frac{P(H_i \cap (E_1 \cap E_2 \cap \cdots \cap E_n))}{P(E_1 \cap E_2 \cap \cdots \cap E_n)} \\
&= \frac{P((E_1 \cap E_2 \cap \cdots \cap E_n) \cap H_i)}{\sum_{j=1}^{m} P((E_1 \cap E_2 \cap \cdots \cap E_n) \cap H_j)} \\
&= \frac{P((E_1 \cap E_2 \cap \cdots \cap E_n)|H_i) \times P(H_i)}{\sum_{j=1}^{m} P((E_1 \cap E_2 \cap \cdots \cap E_n)|H_j) \times P(H_j)}
\end{aligned}$$

where $1 \leq i \leq m$.

The above series of computations are mathematically sound, but the involvement of joint conditional probabilities that have to deal with all possible combinations of evidences and hypotheses is way too complicated for a human expert to conclude from their real life experiences. To make this mathematical rationale more practical and easier to apply in real life, the conditional independence among these evidences is usually assumed so that:

$$\begin{aligned}
&P((E_1 \cap E_2 \cap \cdots \cap E_n)|H_i) \\
&= P(E_1|H_i) \times P(E_2|H_i) \times \cdots \times P(E_n|H_i) \times P(H_i)
\end{aligned}$$

As a result, the mathematical result of this posterior probability can be further simplified into:

$$\begin{aligned}
&P(H_i|(E_1 \cap E_2 \cap \cdots \cap E_n)) \\
&= \frac{P(E_1|H_i) \times P(E_2|H_i) \times \cdots \times P(E_n|H_i) \times P(H_i)}{\sum_{j=1}^{m} P(E_1|H_j) \times P(E_2|H_j) \times \cdots \times P(E_n|H_j) \times P(H_j)}
\end{aligned}$$

where $1 \leq i \leq m$.

This assumption significantly simplifies the complexity of the human's mental burden and makes the inference processing cognitively workable. Instead of considering all evidences and all hypotheses at once, the human experts are now considering an evidence and a hypothesis at a time.[3]

3. The Object-Oriented Design

Object-Oriented approach has been well recognized as a good practice to produce quality reusable software. Although this probabilistic inference process looks specific, it involves intensive calculations between prior probabilities and conditional probabilities to infer the intended posterior probabilities. To design this inference engine as an reusable and general purpose module, four class types are modeled and presented in the following subsections.[4]

3.1. *The Class Type to Model a Prior Probability*

A prior probability within a probabilistic inference domain is an unconditional likelihood that supports a hypothesis without taking any evidence into consideration, such as $P(H_1)$, $P(H_2), \ldots,$ or $P(H_n)$. This class type is named as *PrioProb*. The methods within the *PrioProb* class are listed as follows:

- Constructor:
  The *Constructor* method is used to initialize the prior probability.
- ToString:
  The *ToString* method is used to convert the prior probability into a string representation.

The attributes within the *PrioProb* class type are listed as follows:

- hypo:
  The *hypo* attribute represents the hypothesis in the prior probability.
- prob:
  The *prob* attribute represents the likelihood of the prior probability.

3.2. *The Class Type to Model a Conditional Probability*

A conditional probabilities within a probabilistic inference domain is a conditional likelihood that an evidence will be revealed if a given hypothesis is true, such as $P(E_i|H_1)$, $P(E_i|H_2), \ldots,$ or $P(E_i|H_n)$. This class type is named as *CondProb*. The methods within the *CondProb* class are listed as follows:

- Constructor:
  The *Constructor* method is used to initialize the conditional probability.
- ToString:
  The *ToString* method is used to convert the conditional probability into a string representation.

The attributes within the *CondProb* class type are listed as follows:

- hypo:
  The *hypo* attribute represents the hypothesis in the conditional probability.
- evid:
  The *evid* attribute represents the evidence in the conditional probability.
- prob:
  The *prob* attribute represents the likelihood of the conditional probability.

3.3. *The Class Type to Model a Knowledge Base*

The knowledge base within a probabilistic inference domain consists of a list of prior probabilities and a list of conditional probabilities. This class type is named as *KnowledgeBase*. The methods within the *KnowledgeBase* class are listed as follows:

- Constructor:
  The *Constructor* method is used to initialize the knowledge base.
- ToString:
  The *ToString* method is used to convert the knowledge base into a string representation.

The attributes within the *KnowledgeBase* class type are listed as follows:

- listOfPriorProbs:
  The *listOfPriorProbs* attribute represents the list of prior probabilities in the knowledge base.
- listOfCondProbs:
  The *listOfCondProbs* attribute represents the list of conditional probabilities in the knowledge base.

3.4. *The Class Type to Model a Posterior Probability*

A posterior probability within a probabilistic inference domain is an inferred result that represents the likelihood of a hypothesis to be true if some evidences are shown to be true, such as $P(H_i|(E_1 \cap E_2 \cap ... \cap E_n))$. This class type is named as *PostProb*. The methods within the *PostProb* class are listed as follows:

- Constructor:
  The *Constructor* method is used to initialize the domain knowledge.

- Inference:
  The *Inference* method applies Bayesian inference steps to compute the resultant posterior probability.
- ToString:
  The *ToString* method is used to convert the resultant posterior probability into a string representation.

The attributes within the *PostProb* class type are listed as follows:

- kb:
  The *kb* attribute represents the current knowledge base.
- hypo:
  The *hypo* attribute represents the hypothesis of the posterior probability.
- listOfEvids:
  The *listOfEvids* attribute represents the list of evidences that have been shown to be true.

4. The Implementation in Python Programming Language

In light of the abundant built-in data types as well as the prevalence in real life problem solving, this inference engine consists of the aforementioned four class types is implemented in Python programming language. The source code turns out to be a reusable module. This module is lightweight, domain independent and can be plugged into future probabilistic inference applications to speed up their development.[4]

4.1. *The Source Code of the Inference Engine in Python Programming Language*

```
"""
File Name: bayesian.py
Creator: Feng-Jen Yang
Purpose: Implementing a Bayesian inference engine
"""

class PrioProb(object):
    """Representing a prior probability"""

    def __init__(self, initHypo, initProb):
        """Initializing the prior probability"""
        self.hypo = initHypo
        self.prob = initProb
```

```
    def __str__(self):
        """Returning a string representation of the prior
        probability"""
        return "P(" + self.hypo + ") = " + str(self.prob)

class CondProb(object):
    """Representing a conditional probability"""

    def __init__(self, initEvid, initHypo, initProb):
        """Initializing the conditional probability"""
        self.evid = initEvid
        self.hypo = initHypo
        self.prob = initProb

    def __str__(self):
        """Returning a string representation of the conditional
        probability"""
        return "P(" + self.evid + "|" + self.hypo + ") = " +
        str(self.prob)

class KnowledgeBase(object):
    """Representing a knowledge domain"""

    def __init__(self, initPrioProbs, initCondProbs):
        """Initializing the prior probabilities and conditional
        probabilities"""
        self.listOfPrioProbs = initPrioProbs
        self.listOfCondProbs = initCondProbs

    def __str__(self):
        """Returning a string representation of the knowledge base"""
        result = "The Knowledge Base:"
        for p in self.listOfPrioProbs:
            result += "\n" + str(p)
        for p in self.listOfCondProbs:
            result += "\n" + str(p)
        return result

class PostProb(object):
    """Representing Bayesian inference"""
    def __init__(self, initKB, initHypo, initEvids):
        """Initializing the knowledge base"""
        self.kb = initKB
        self.hypo = initHypo
        self.listOfEvids = initEvids
```

```
def inference(self):
    """Computing the posterior probability"""
    numerator = 1
    for cp in self.kb.listOfCondProbs:
        if (cp.hypo == self.hypo) and (cp.evid in self.
        listOfEvids):
            numerator *=cp.prob
    for pp in self.kb.listOfPrioProbs:
        if pp.hypo == self.hypo:
            numerator *= pp.prob
            break

    denominator = 0
    for pp in self.kb.listOfPrioProbs:
        term = 1
        for cp in self.kb.listOfCondProbs:
            if (cp.hypo == pp.hypo) and (cp.evid in self.
            listOfEvids):
                term *=cp.prob
        term *= pp.prob
        denominator += term
    return round(numerator/denominator, 3)

def __str__(self):
    """Returning a string representation of the inference
    reault"""
    result = "The Posterior Probability:\nP(" + self.hypo
    + "|("
    for e in self.listOfEvids:
        result += e

        if e != self.listOfEvids[-1]:
            result += "^"
    result +=  ")) = " + str(self.inference())
    return result
```

5. Applied in Real Life Probabilistic Inference

This module has been well tested to be domain independent and easy to be imported into a python program in which it can be instantiated to fit a specialized knowledge domain for further purposeful inference. For the purpose of demonstration, a sample application of applying this inference engine are described in the following subsections.

5.1. *A Sample Application of Using this Inference Engine*

Assume that we are provided with a medical database by which we summarized the following prior probabilities and conditional probabilities:

- At any point in time, 0.01% of people have cold.
- 0.88% of people who are having cold will have fever.
- 0.78% of people who are having cold will cough.
- 0.68% of people who are having cold will sneeze.

Based on the above the information, the following three sample posterior probabilities can be inferred by importing this inference engine as a module into a Python program:

- If a person is having fever and sneeze, how much likely is this person having a cold?
- If a person is having cough and sneeze, how much likely is this person not having a cold?
- If a person is having fever, cough and sneeze, how much likely is this person having a cold?

As shown in the program output subsection, the resultant posterior probabilities of these three inferred questions are:

- $P(Cold|(Fever \cap Sneeze) = 0.006$
- $P(NoCold|(Cough \cap Sneeze) = 0.9954$
- $P(Cold|(Fever \cap Cough \cap Sneeze) = 0.005$

The application program that instantiates the above knowledge domain and performs the intended inferences are illustrated in the next two subsections.

5.2. *The Sample Application Program in Python Program Language*

```
"""
File Name: demo.py
Creator: Feng-Jen Yang
Purpose: A demonstration of using this Bayesian inference engine
"""

from bayesian import *

def main():
```

```
    """The main function that coordinates the program execution"""

    #Instantiate the prior probabilities
    listOfPrioProbs = [PrioProb("Cold", 0.01), PrioProb("NoCold",
    0.99)]

    #Instantiate conditional probabilities
    listOfCondProbs = [CondProb("Fever", "Cold", 0.88),\
                       CondProb("Cough", "Cold", 0.78),\
                       CondProb("Sneeze", "Cold", 0.68)]

    #Instantiate the knowledge base
    kb = KnowledgeBase(listOfPrioProbs, listOfCondProbs)

    #Display the knowledge base
    print(kb, "\n")

    #Perform the Bayesian inference to compute the posterior
    probability
    pp = PostProb(kb, "Cold", ["Fever", "Sneeze"])

    #display the inference result
    print(pp, "\n")

    #Perform the Bayesian inference to compute the posterior
    probability
    pp = PostProb(kb, "NoCold", ["Cough", "Sneeze"])

    #display the inference result
    print(pp, "\n")

    #Perform the Bayesian inference to compute the posterior
    probability
    pp = PostProb(kb, "Cold", ["Fever", "Cough", "Sneeze"])

    #display the inference result
    print(pp, "\n")

#The entry point of program execution
main()
```

5.3. *The Output from the Sample Program*

```
The Knowledge Base:
P(Cold) = 0.01
P(NoCold) = 0.99
```

```
P(Fever|Cold) = 0.88
P(Cough|Cold) = 0.78
P(Sneeze|Cold) = 0.68

The Posterior Probability:
P(Cold|(Fever^Sneeze)) = 0.006

The Posterior Probability:
P(NoCold|(Cough^Sneeze)) = 0.995

The Posterior Probability:
P(Cold|(Fever^Cough^Sneeze)) = 0.005
```

6. Conclusion and Future Work

Bayesian inference can be easily applied as a machine learning method, such as optimal Bayesian classier, naive Bayesian classifier,[5] and collaborative filtering.[6] This inference engine is based on the Bayesian theory and implemented in Python programming language. As a performance-centered design this inference engine is functioning comprehensively without consuming excessive computational resources. I am aiming at publishing this inference engine as an open source to further benefit both industrial application developments and academic researches. A similar project to create a domain independent software module for Bayesian belief network is being planned. Instead of performing probabilistic inference for posterior probabilities, a Bayesian belief network module will perform the probabilistic estimations for disjoint probabilistic distributions.

References

1. T. Bayes, *An Essay Towards Solving a Problem in the Doctrine of Chances*, (Philosophical Transactions of the Royal Society), Volume 53, Issue 1,1763, pp. 370-418.
2. J. Tabak, *Probability and Statistics: The Science of Uncertainty*, (NY: Facts On File, Inc., USA, 2004), pp. 46-50.
3. Feng-Jen Yang, *Eliciting an Overlooked Aspect of Bayesian Reasoning*, (*ACM SIGCSE* Bulletin), Volume 39, Issue 4, 2007, pp. 45-48.
4. *Feng-Jen Yang, "Crafting a Lightweight Bayesian Inference Engine," Lecture Notes in Engineering and Computer Science: Proceedings of The World Congress on Engineering 2016, 29 June-1 July, 2016, London, U.K.*, pp. 612-617.

5. A. Gelman, D. Rubin and H. S. Stern, *Bayesian Data Analysis* (CRC Press 2003).
6. J. S. Breese, D. Heckerman and C. Kadie, *Empirical Analysis of Predictive Algorithms for Collaborative Filtering*, (In Proceedings of the Fourteenth Conference on Uncertainty in Artificial Intelligence, Madison, WI), 1998, pp. 43-52.

CONSTRUCTION OF ANALYTICAL SOLUTIONS TO THE BLACK-SCHOLES OPTION VALUATION MODEL BY MEANS OF HE'S POLYNOMIALS TECHNIQUE*

S. O. EDEKI[†,§], *Member, IAENG,* O. O. UGBEBOR[†,‡,¶] and E. A. OWOLOKO[†,||]
[†]*Department of Mathematics, Covenant University,*
[‡]*Department of Mathematics, University of Ibadan, Ibadan, Nigeria*
[§]*soedeki@yahoo.com*
[¶]*ugbebor1@yahoo.com*
[||]*enahoro.owoloko@covenantuniversity.edu.ng*

In this paper, a semi-analytical method referred to as He's Polynomials Technique (HPT) is proposed for solving the classical Black-Scholes pricing model with stock as the underlying asset. The proposed method gives the exact solution of the solved problem in a very simple and quick manner even with less computational work while still maintaining high level of accuracy. Hence, we recommend an extension and adoption of this method for solving problems arising in other areas of financial engineering, finance, and applied sciences.

Keywords: Analytical solutions; Black-Scholes model; He's polynomials; stock option valuation.

1. Introduction

The Black-Scholes model is one of the most famous and useful models for option valuation as regards option pricing theory. Financial engineering appears to be among the most and famous areas of applied mathematics from both theoretical, industrial and academic point of view. In the theory of finance, and options valuation; options are referred to as financial contracts furnishing its holder the right but not an obligation to be involved in a future transaction on some underlying securities [1]. Therefore, an option having stock as its underlying asset is referred to as stock option. One of the most fundamental aspects of financial mathematics is hinged on modelling of evolution of financial processes such as interest rates, rates of exchange, stock prices, and so on [2-4].

* Work partially supported by CUCRID: Covenant University Centre for Research, Innovation and Discovery.

In order to control the risk induced by the movements of stock prices, options can be used for hedging assets and portfolios. With regard to theory of option pricing and valuation, Black and Scholes in 1973 [5] proposed a classical formula for the prices of financial options. This is popularly referred to as Black-Scholes equation, which has been the hallmark of financial derivatives. The Black-Scholes model is a linear PDE based on some assumptions whose modification and relaxation have also led to various forms with nonlinear cases inclusive [6,7].

2. The Black-Scholes Model

In financial engineering, the Black-Scholes model is a parabolic differential equation whose solution is used in describing the value of European option [8]. In these recent days, the Black-Scholes model has been extended for valuation of options other than European options.

In what follows, we will consider the classical celebrated Black–Scholes option pricing model:

$$\frac{\partial f}{\partial \tau}+\frac{1}{2}S^2\sigma^2\frac{\partial^2 f}{\partial S^2}+rS\frac{\partial f}{\partial S}-rf=0 \tag{1}$$

where $f=f\left(S,\tau\right)$ is the value of the contingent claim S, at time, τ $0\leq\tau\leq t,\ \left(S,\tau\right)\in R^+\times\left(0,T\right), f\in C^{2,1}\left[R\times[0,T]\right]$ with a payoff function $p_f\left(S,t\right)$, and expiration price, E such that:

$$p_f\left(S,t\right)=\begin{cases}\max(S-E,0), & \text{for European call option}\\ \max(E-S,0), & \text{for European put option}\end{cases} \tag{2}$$

where $\max\left(S_*,0\right)$ indicates the large value between S_* and 0, σ denotes the volatility of the underlying asset $S=S(t)$, r is the risk-free interest rate, T is the time at maturity. Note, the randomness of $f=f(S,t)$ will be totally correlated to that of $S=S(t)$.

Recently, many researchers have considered some numerical, and analytical methods for solving the Black–Scholes model, and other option valuation models. These include Adomian decomposition method (ADM), variational iteration method (VIM), Differential transformation method (DTM), projected DTM (PDTM), homotopy perturbation method (HPM), [9-15].

The He's polynomials was introduced by Ghorbani *et al.* [16, 17] where nonlinear terms were split into a series of polynomials which are calculated from HPM. It is remarked that He's polynomials are compatible with Adomian's polynomials, yet it is showed that the He's polynomials are easier to compute, and are very much user friendly [18-22].

Our goal is therefore, to extend the conference paper of Edeki *et al.* [23] by constructing analytical solutions to the Black–Scholes option pricing model of the form in (1) via a semi-analytical technique known as He's polynomials technique. It is worth mentioning that the He's polynomials method is an alternative semi-analytical method, even without giving up accuracy.

3. Overview of the He's Polynomial Method

In general form, we consider the equation:

$$\Im(\psi)=0 \tag{3}$$

where $\Im$ is an integral or a differential operator. Let $H(\psi,p)$ be a convex homotopy defined by:

$$H(\psi,p)=p\Im(\psi)+(1-p)G(\psi) \tag{4}$$

where $G(\psi)$ is a functional operator with ψ_0 as a known solution. Thus, we have:

$$H(\psi,0)=G(\psi) \text{ and } H(\psi,1)=\Im(\psi) \tag{5}$$

whenever $H(\psi,p)=0$ is satisfied, and $p\in(0,1]$ is an embedded parameter. In HPM [18,19], p is used as an expanding parameter to obtain:

$$\psi=\sum_{j=0}^{\infty}p^j\psi_j=\psi_0+p\psi_1+p^2\psi_2+\cdots \tag{6}$$

From (6) the solution is obtained as $p\to 1$. The convergence of (9) as $p\to 1$ has been considered in [20].

The method considers $N(\psi)$ (the nonlinear term) as:

$$N(\psi)=\sum_{j=0}^{\infty}p^jH_j=H_o+p^1H_1+p^2H_2+p^3H_3+\cdots \tag{7}$$

where H_k's are the so-called He's polynomials, which can be computed using:

$$H_k=\frac{1}{k!}\frac{\partial^k}{\partial p^k}\left(N\left(\sum_{j=0}^{k}p^j\psi_j\right)\right)_{p=0},\ n\geq 0 \tag{8}$$

where $H_k=H_k(\psi_0,\psi_1,\psi_2,\psi_3,\ldots,\psi_k)$

4. The He's Polynomials and the Pricing Model

Aaa In this subsection, the He's Polynomials approach will be applied to the model equation as follows:

Problem 1: Consider the following linear Black-Scholes equation:

$$\frac{\partial w}{\partial t}+x^2\frac{\partial^2 w}{\partial x^2}+\frac{1}{2}x\frac{\partial w}{\partial x}-w=0 \tag{9}$$

subject to:

$$w(x,0)=\max(x^3,0)=\begin{cases} x^3, & \text{for } x>0 \\ 0, & \text{for } x\leq 0 \end{cases} \tag{10}$$

Procedure w.r.t Problem 1:

We re-write (9) as (11) and in an integral form, we reformulate (11) as (12) below:

$$\frac{\partial w}{\partial t}=-x^2\frac{\partial^2 w}{\partial x^2}-\frac{1}{2}x\frac{\partial w}{\partial x}+w \tag{11}$$

$$w(x,t)=\max\left(x^3,0\right)-p\int_0^t\left\{x^2\frac{\partial^2 w}{\partial x^2}+\frac{1}{2}x\frac{\partial w}{\partial x}-w\right\}dt \tag{12}$$

Applying the convex homotopy method to (12) gives:

$$\sum_{n=0}^{\infty}p^n w_n=\max\left(x^3,0\right)-p\int_0^t\left\{x^2\sum_{n=0}^{\infty}p^n\frac{\partial^2 w_n}{\partial x^2}+\frac{x}{2}\sum_{n=0}^{\infty}p^n\frac{\partial w_n}{\partial x}-\sum_{n=0}^{\infty}p^n w_n\right\}dt \tag{13}$$

$\Rightarrow$

$$\begin{aligned} w_0+pw_1+p^2w_2+\cdots=\max\left(x^3,0\right)-p\int_0^t\Bigg\{&x^2\left(\frac{\partial^2 w_0}{\partial x^2}+p\frac{\partial^2 w_1}{\partial x^2}+p^2\frac{\partial^2 w_2}{\partial x^2}+\cdots\right)\\ &+\frac{x}{2}\left(\frac{\partial w_0}{\partial x}+p\frac{\partial w_1}{\partial x}+p^2\frac{\partial w_2}{\partial x}+\cdots\right)\\ &-(w_0+pw_1+p^2w_2+\cdots)\Bigg\}dt \end{aligned} \tag{14}$$

We compare the coefficients of the equal powers of p in the following ways:

$$p^{(0)}:w_0=\max\left(x^3,0\right)$$

$$p^{(1)}:w_1=-\int_0^t\left(x^2\frac{\partial^2 w_0}{\partial x^2}+\frac{x}{2}\frac{\partial w_0}{\partial x}-w_0\right)dt$$

$$p^{(2)}:w_2=-\int_0^t\left(x^2\frac{\partial^2 w_1}{\partial x^2}+\frac{x}{2}\frac{\partial w_1}{\partial x}-w_1\right)dt$$

$$p^{(3)}: w_3 = -\int_0^t \left(x^2 \frac{\partial^2 w_2}{\partial x^2} + \frac{x}{2} \frac{\partial w_2}{\partial x} - w_2 \right) dt$$

$$\vdots$$

$$p^{(m)}: w_m = -\int_0^t \left(x^2 \frac{\partial^2 w_{m-1}}{\partial x^2} + \frac{x}{2} \frac{\partial w_{m-1}}{\partial x} - w_{m-1} \right) dt$$

Therefore, simplifying $p^{(1)}$, $p^{(2)}$, $p^{(3)}$, and so on using $w_0 = \max\left(x^3, 0\right)$, for $x \geq 0$ gives:

$$w_0 = x^3,\ w_1 = -\frac{13}{2} x^3 t,\ w_2 = \frac{169}{8} x^3 t^2,\ w_3 = \frac{-2197}{48} x^3 t^3,$$

$$w_4 = \frac{28561}{384} x^3 t^4,\ w_5 = \frac{-371293}{3840} x^3 t^5, \cdots$$

$$\therefore\ w(x,t) = w_0 + w_1 + w_2 + w_3 + w_4 + w_5 + \cdots$$

$$= x^3 - \frac{13}{2} x^3 t + \frac{169}{8} x^3 t^2 - \frac{2197}{48} x^3 t^3 + \frac{28561}{384} x^3 t^4 - \frac{371293}{3840} x^3 t^5 + \cdots$$

$$= x^3 \left(1 + (-6.5t) + \frac{(-6.5t)^2}{2!} + \frac{(-6.5t)^3}{3!} + \frac{(-6.5t)^4}{4!} + \frac{(-6.5t)^5}{5!} + \cdots \right)$$

$$\cong x^3 e^{-6.5t} \tag{15}$$

Note: Figures 1 and 2 represent the graphs of the exact solution and the He's polynomials solution (including terms up to power 5) respectively:

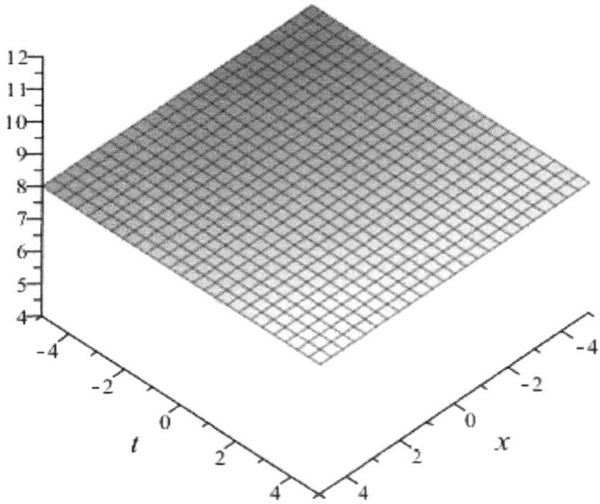

Fig. 1. Graphical view of the exact solution.

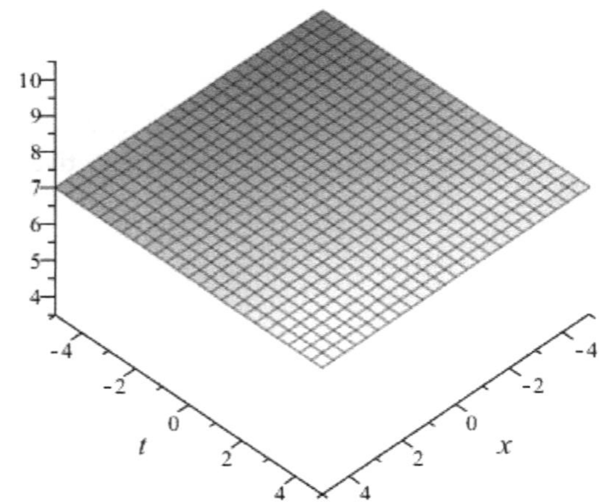

Fig. 2. Graphical view of the approximate solution.

5. Concluding Remarks

We have succeeded in providing accurate and exact solutions to the classical linear Black-Scholes model for stock option valuation. This is done by the means of He's polynomials as the proposed technique. This technique is very much efficient and reliable as it gives the exact solution of the solved problem in a very simple and quick manner even with less computational work while still maintaining high level of accuracy. Hence, we recommend an extension and adoption of this method for solving problems arising in other areas of applied sciences: financial engineering and finance. We used *Maple* 18 for all numerical computations, and graphics done in this study.

Acknowledgments

The authors wish to thank Covenant University Centre for Research, Innovation and Discovery (CUCRID) for supporting this work financially. Also, all constructive and helpful remarks from the anonymous referee(s) are highly appreciated.

References

1. J. C. Hull, "Options, Futures, and Other Derivatives", *New Jersey: Prentice-Hall,* 1999.
2. F. Mehrdoust and M. Mirzazadeh, "On Analytical Solution of the Black-Scholes Equation by the First Integral Method", *UPB Scientific Bulletin, Series A,* **76** (4), (2014): 85-90.
3. M.E. Adeosun, S.O. Edeki and O.O. Ugbebor, "Stochastic Analysis of Stock Market Price Models: A Case Study of the Nigerian Stock Exchange (NSE)", *WSEAS Transactions on Mathematics,* **14** (2015): 353-363.
4. O.O. Ugbebor and S.O. Edeki, "On Duality Principle in Exponentially Levy Market", *Journal of Applied Mathematics & Bioinformatics,* **3** (2), (2013), 159-170.
5. F. Black and M. Scholes, "The Pricing of Options and Corporate Liabilities'', *Journal of Political Economy,* **81** (1973) 637-654.
6. S.O. Edeki, O.O. Ugbebor and E.A. Owoloko, "On a Dividend-Paying Stock Options Pricing Model (SOPM) Using Constant Elasticity of Variance Stochastic Dynamics", *International Journal of Pure and Applied Mathematics,* **106** (4), (2016): 1029-1036.

7. O. González-Gaxiola, J. R. de Chávez and J. A. Santiago, "A Nonlinear Option Pricing Model Through the Adomian Decomposition Method", *International Journal of Applied and Computational Mathematics*, DOI 10.1007/s40819-015-0070-6: (2015).
8. R.C. Merton, "Theory of Rational Options Pricing", *The Bell Journal of Economics and Management Science*, **4** (1), (1973): 141-183.
9. V. Gülkaç, "The Homotopy Perturbation Method for the Black–Scholes Equation", *Journal of Statistical Computation and Simulation* **80** (12), (2010): 1349–1354.
10. W. MacMillan, "An Analytical Approximation for the American Put Prices", *Adv. Futures Options Res.* **1** (1986): 119-139.
11. G. Barone-Adesi and R. Elliot, "Approximations for the Values of American Options", Stochastic, *Analysis and Applications*, **9** (1991): 115–131.
12. J. Esekon, S. Onyango and N.O. Ongati, "Analytic Solution of a Nonlinear Black-Scholes Partial Differential Equation", *International Journal of Pure and Applied Mathematics,* **61** (2), (2010): 219-230.
13. J. Ankudinova and M. Ehrhardt, "On the Numerical Solution of Nonlinear Black-Scholes Equations", *Computers and Mathematics with Applications*, **56**, (2008): 799–812.
14. H. Aminikhah and F. Mehrdoust, "On Approximate-analytical Solution of Generalized Black-Scholes Equation", *UPB Scientific Bulletin, Series A,* **77** (4), (2015): 186-194.
15. S.O. Edeki, I. Adinya and O.O. Ugbebor, "The Effect of Stochastic Capital Reserve on Actuarial Risk Analysis via an Integro-differential Equation" *IAENG International Journal of Applied Mathematics*, **44** (2), (2014): 83-90.
16. A. Ghorbani and J. S. Nadjfi, "He's Homotopy Perturbation Method for Calculating Adomian's Polynomials", *International Journal of Nonlinear Sciences and Numerical Simulation*, **8** (2) (2007), pp. 229-332.
17. A. Ghorbani, "Beyond Adomian's Polynomials: He polynomials", *Chaos, Solitons & Fractals,* (2007), in press.
18. J.H. He, "Homotopy Perturbation Method: A New Nonlinear Analytical Technique", *Journal of Applied Mathematics and Computation*, **135** (2003), 73-79.
19. S.O. Edeki, G.O. Akinlabi and M.E. Adeosun, "Analytical Solutions of the Navier-Stokes Model by He's Polynomials, " Lecture Notes in Engineering and Computer Science: *Proceedings of the World Congress on Engineering* 2016, WCE 2016, June 29 - July 1, 2016, London, U.K., pp. 16-19.
20. S.T. Mohyud-Din, "Solution of Troesch's Problem using He's Polynomials", *Revista De La Union Matematica Argentina,* **52** (1), (2011): 143-148.

21. A. El-Ajou, Z. Odibat , S. Momani and A. Alawneh, "Construction of Analytical Solutions to Fractional Differential Equations Using Homotopy Analysis Method", *IAENG International Journal of Applied Mathematics*, **40**: 2, 43-51, 2010.
22. A. Cheniguel and M. Reghioua, "Homotopy Perturbation Method for Solving Some Initial Boundary Value Problems with Non Local Conditions," Lecture Notes in Engineering and Computer Science: *Proceedings of the World Congress on Engineering and Computer Science* 2013, WCECS 2013, 23-25 October, 2013, San Francisco, USA, pp. 572-577.
23. S.O. Edeki, O.O. Ugbebor and E.A. Owoloko, "He's Polynomials for Analytical Solutions of the Black-Scholes Pricing Model for Stock Option Valuation," Lecture Notes in Engineering and Computer Science: *Proceedings of the World Congress on Engineering* 2016, WCE 2016, June 29 - July 1, 2016, London, UK, pp. 632-634.

A COLLABORATIVE SOFTWARE DEVELOPING METHOD BY MIMICKING ARMY ANT ALTRUISM BEHAVIORS

T. ICHIMURA* and T. UEMOTO

Prefectural University of Hiroshima
Hiroshima, 733-0834, Japan
**ichimura@ieee.org*
http://www.pu-hiroshima.ac.jp/

S. KAMADA

Graduate School of Information Sciences, Hiroshima City University
Hiroshima, 731-3194, Japan
da65002@e.hiroshima-cu.ac.jp

Recently, social collaboration that helps more people or groups interact and share information to reach their common goal has been receiving a lot of attention. In the process of developing software, the social collaboration is required because they are complex tasks where many projects are processed and many developers join two or more projects simultaneously. To improve the efficiency of developing software, we focus the activities of developers in a group. The community of the developers is analyzed by analyzing method of network structure. We develop the simulation system of developing software with altruism behaviors. The system was designated on the basis of developers behaviors of SourceForge.net. We observed the community growing process under the agent's altruism behaviors and can realize the altruism behavior in the software developing community.

Keywords: Altruism behavior; Analysis of Group development Process; Growing Process of Social Community

1. Introduction

Recently, social collaboration is receiving a lot of attention because it argues the process that helps more people or groups interact and share information to reach their common goal. A social community is composed of a group of individuals with the common interest and purpose. The size of community is determined by the number of people and their activities. We can consider that the attraction to the community is increased according to the integration of the interaction of the personal behaviors, and then it grows to be a larger society.

In the community of open source software (OSS) developers and projects, we can observe the collaborative social network in the developing process of system. Madey *et al.*[1] has been developed the structural and the dynamic mechanisms that govern the topology and the evolution of OSS developer's social community to analyze the developer and project data from Source Forge.[4] According to their ideas, the collaborative social network is formed in global virtual self-organizing teams of software developers working on OSS projects and they developed the self-organizing simulation model. Because the self-organization mechanism is the important property of many social networks, a self-organizing network grows and develops the structure without centralized decision making as shown in the real OSS community.

In their model, the developer in the OSS community is an agent. Some agents create a new project (community) together. The project is the team to develop a software (system). The agent can select only an action representing a real developer's possible daily interaction in the process of developing system. The model described in Ref. 2 assumed that the project has some tasks and the agent has the skill for the developing software. The 'task' is the amount of work for the developing software. The 'skill' is the amount of works that the agent develops the program for a given time. Our developed agent based simulation model shows some spiral line in the relation between the community and the action of agents.

In this paper, we propose the collaborative software developing method by mimicking altruism behaviors of army ant agent system.[3] The agent decides to join into the project on the based on the altruism behaviors by imitating the army ants. If the agent with high ability has some working time to find the project without a lot of efforts, the agent leads to participate in the developing team. Their tasks will be increased, but the total efficiency of developing software will be increased. This paper discusses the relation between the number of altruism agents and total cost of projects through some experiments by using simulation system for the Source Forge community data.[4]

2. Open Source Community Data

There are some software distribution web sites for OSS communities such as GNU project. Source Forge is one of popular software repositories and has the function of the management system of open source software. The website of Source Forge gives the information of the projects and their developers recorded in the database.[5] The data related to project and developer are obtained from the database. We can investigate the relation between the number of projects and developers who join the project.

Fig.1 shows the logarithm of the number of projects and developers at September 2014 (Schema Sf0914). Schema Sf0914 has 73 tables including 'User Group.' The table of 'User Group' consists of 17 columns including 'user_id' and 'group_id'. We retrieved all data from 'sf0914.user_group' and counted the records according to 'group_id'. As a result, the number of developers belonged to each project is calculated. The horizontal axis is the number of developers and the vertical axis is the number of projects. As shown in Fig. 1, we can observe that the relation is decreased linearly for a certain period and then the line spreads from the certain point like delta. In this paper, we investigate the inflection point and the expanse of the delta, and the equation of line is determined according to empirical studies.

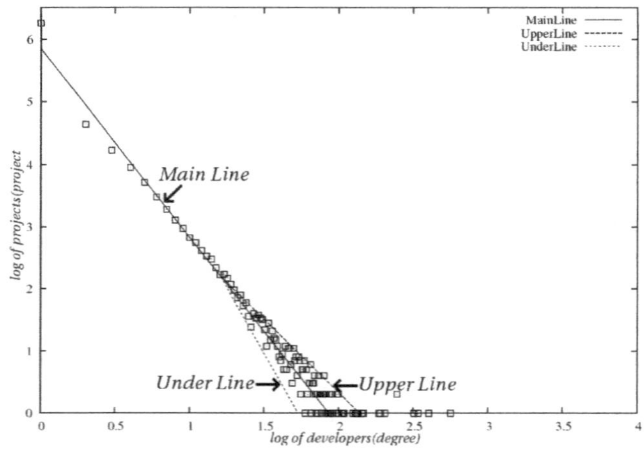

Fig. 1. The distribution for project and developer in Source Forge.

3. Simulation Model

In our developed simulation system, there are 2 kinds of agents: 'major agent' and 'minor agent'. The agent has the own function of creating, joining to, or leaving from the project. Because some developers become to show the leadership in the real world of software development, the simulation model has the 2 kinds of agents behaviors inspired by the interactive operation of army ant system.[6,7]

In our model, a major agent creates the project, and a minor agent takes the participating behavior to the project and the leaving behavior from one. A major agent and a minor agent is a project manager and a developer, respectively. When some minor agents join the same project, they worked to share the task of software development. The 'task' and the 'skill' means the numerical values representing the amount of codes for the software development and the agent's coding capability in the periodical time, respectively. Our developed agent based simulation model shows some spiral lines in the relation between the community and the action of agents as shown in Fig. 1.

Source Forge can provide open source softwares which are classified into 10 categories such as "Audio & Video," "Business & Enterprise," "Communications," "Development," "Home & Education," "Games," "Graphics," "Science & Engineering," "Security & Utilities," and "System Administration."[4] If a major agent creates the project, it must determine the project's category. The development work will have two or more types of tasks during developing the software. There are 3 kinds of common types of tasks; 'network communication processing,' 'database processing,' and 'graphic processing.' The minor agent cannot join the project, if it does not have the skill for the corresponding category. If two or more agents join a project, the task for the developing project is shared to each agent equally.

3.1. *Task and Skill*

A project has one or more tasks in the software development. Let P_i^k be the amount of task k in the project i. P_i^k takes a value in $[0, 2]$. The vector of P_i^k for all tasks is defined as $\mathbf{P_i} = \{P_i^1, P_i^2, \ldots, P_i^n\}$.

A skill for the agent j in task k is defined as S_j^k. The initial value is a given random number in $[0, 1]$. The agent works the allocated task with the agent's skill S_j^k. That is, the time to finish the task is calculated by dividing the amount of task of an agent W_{ij}^k by the agent skill S_j^k. The working time for each project to which the agent joins is added. The total time T_j is the summation of T_{ij} for each agent as follows:

$$T_j = \sum_i T_{ij} = \sum_i \frac{W_{ij}^k}{S_j^k}, \tag{1}$$

where W_{ij}^k is the amount of the task k which is assigned to the agent j in the project i. That is, $W_{ij}^k = \frac{P_i^k}{N_i}$. N_i is the number of agents in the project *i*. T_j satisfies the condition $0 \leq T_j \leq T_{limit}$. The upper limit of time T_{limit} is 24 hours a day.

3.2. *The behaviors of agent*

In Ref. 1, the proposed model has 4 kinds of behaviors: "Developer creates a project (create a project)," "Developer joins into a project (join a project)," "Developer abandons a project (leave a project)" or "Developer continues the current collaboration (no any actions)." By such actions, a new project can be created and be grown, although a few projects are disappeared. The agent can contact to other agent in another community and join two or more projects simultaneously. The change in the growing community can be observed through the simulation. The experimental result described in Ref. 1 showed the linear relation between the number of projects and developers. In this paper, the model has 2 kinds of agents: Major agent and Minor agent. A Major agent works to create a project. Minor agents has 3 kind of properties such as 'Join a project,' 'Leave a project,' and 'No action.' A Minor agent selects an action from 3 kinds of behaviors per a step and it works according to the selected behavior by the agent's action algorithm.

3.3. *Create a project*

A Major agent creates a new project when it starts to develop a new software. In this case, a created project is called New one. Moreover, an agent creates a new one, when the agent creates sub projects to share the large tasks in the project which is created by the Major agent. In this case, a created project is called Sub one. If a Major agent j creates a new project, the agent works to follow next behavior's algorithm.

(1) The probabilities New or Sub of creating a project for each case are given appropriate values.
(2) If a Major agent creates a new project, an agent checks the condition that the probability New equals to or greater than the random value $[0, 1]$ by using Monte Carlo method.
(3) If a Major agent creates a sub project to share the tasks, the Major agent determines according to Eq. (2) and Eq. (3).

$$Sub_i^k = \frac{P_i^k}{N_i^k}, \tag{2}$$

where Sub_i^k is the average amount of tasks of a Minor agent in the developing task k of the project i. P_i^k is the task k assigned to the agent and

N_i^k is the number of agents in the project i.

$$\begin{cases} IF\ \ Sub_i^k \geq Sub_{threshold}\ \ THEN\ CreateProject \\ IF\ \ Sub_i^k < Sub_{threshold}\ \ THEN\ NotCreate, \end{cases} \tag{3}$$

where $Sub_{threshold}$ is a given value to share the tasks and is a threshold to create a sub project.

The Major agent checks the value of Sub_i^k for each task which should be shared in the point of load balance. Once a sub project is created, some well-skilled Minor agents in the category of developing task will participate to the sub project.

3.4. *Join a project*

The Minor agent j joins in the project according to the following process.

(1) The Minor agent j selects one project which it has not participated yet.
(2) The probability to be joined in the selected project is determined by Eq. (5).
(3) The Minor agent determines to join in the project by Eq. (6).

If the estimated time till the end of the task exceeds the threshold T_{limit}, the Minor agent gives up the joining in the project.

3.4.1. *Selection of project with Network growing model*

When the agent selects a project, some strategies to a network growing model are required. In this paper, Barabasi-Albert (BA) model with dynamic fitness[8] is adapted to decide the behaviors of agents. BA model is an algorithm for generating random scale-free networks based on the preferential attachment mechanism. Scale-free network is widely used in natural and human-made systems such as social community. In BA model, an edge is connected to the node with high number of degrees in high probability.[9] BA model with dynamic fitness employs the fitness of node is dynamically changed.[8] A degree of software is the number of agents worked in the software development and a Minor agent selects a software based on this model. In this model, the linkage to the other agent is implemented by using BA model, but a new node is not added.

The selection probability S_i for the software i by a Minor agent is defined as follows.

$$S_i = \frac{\eta_i N_i}{\sum \eta_i N_i}, \tag{4}$$

where η_i is the time decreasing function which shows the fitness of software i. N_i is the degree of software, that is, the number of participating agents. The dynamic fitness function realizes the growing community, the stability and the attenuation of community growing.

3.4.2. *Probability of joining a project*

The probability of participation to a project is calculated by the tasks of the project and the skill of the agent j. If the agent does not have enough skills for the category of the project i, the probability is 0. Otherwise, the probability is calculated as follows.

$$J_{ij} = \frac{\sum_{k=1}^{n}(P_i^k - \sum_{j=1} S_j^k)}{\sum_{k=1}^{n} P_i^k} \tag{5}$$

$$\begin{cases} IF \;\; J_{ij} \geq J_{threshold} \;\; THEN \;\; JointProject \\ IF \;\; J_{ij} < J_{threshold} \;\; THEN \;\; NotJoin, \end{cases} \tag{6}$$

where $\sum_{k=1}^{n}(P_i^k - \sum_{j=1} S_j^k)$ is how tasks of the project have reached the completed work, J_{ij} is calculated by dividing the numerator in Eq. (5) by the total amount of tasks $\sum_{k=1}^{n} P_i^k$. If J_{ij} is larger than threshold $J_{threshold}$, a Minor agent joins a project.

3.5. *Leave a project*

The Minor agents can leave from the working project by the following algorithm.

(1) The Minor agent j selects one project among all projects to participate.
(2) The probability of leaving to participate the project is calculated.
(3) The Minor agent decides the project to leave by Eq. (7).

The probability of leaving a project is calculated as follows:

$$L_{ij} = \sum_{k=1}^{n} \left(\frac{P_i^k}{N_i^k} - S_j^k \right) \tag{7}$$

$$\begin{cases} IF \;\; L_{ij} \geq L_{threshold} \;\; THEN \;\; LeaveProject \\ IF \;\; L_{ij} < L_{threshold} \;\; THEN \;\; NotLeave, \end{cases} \tag{8}$$

where $\frac{P_i^k}{N_i^k}$ is the amount of task per an agent, L_{ij} is calculated by subtracting the agent skill from $\frac{P_i^k}{N_i^k}$, and $L_{threshold}$ is a given threshold value. If the load of the agent j (L_{ij}) is greater than $L_{threshold}$, the agent j leaves from a project i.

In Eq. (7), the Minor agent can leave from a project with ease, when the amount of the task of the project is high or the skill of agent is low or in both situations.

4. Agent Behavior with Altruism Behavior

We introduce to focus a mimic model of army ant behaviors in this section.

4.1. *Altruism Behavior in Army Ant Systems*

In animal societies, self-organization is the theory of how minimal complexity in the individual can generate greater complexity at the population. The rules specifying the interactions among the components in the system are implemented by using only local information without global information. In the study of social evolution, army ant performs altruism as one behavior of complexities, where each individual reduces its own fitness but increases the total fitness of other individuals in the population. Such behaviors seem to be involved acts of self-sacrifice in order to aid the others. In evolutionary biology, such a behavior is called reciprocal altruism. The concept was initially developed to explain the evolution of cooperation as mutually altruistic acts.[10]

By inspiring Ichimura *et al.* Ref. 7 developed the multi-agent simulation system to execute more realistic altruistic behavior where two or more kinds of agents realize the division of roles in army ants.[7] Such behaviors are also seen in the collaborative social networks. Such observations of behaviors in the computer simulation strongly will contribute to the shift to knowledge and power from the individual to the collective. We will explore how agent-based modeling and simulation can be used as a research technique to study collaborative social networks.[2,7]

4.2. *Decision Rule for Altruism Behavior*

If the agent participates in the project, the agent makes a link to the other project which it has already joined. The link means the relation among agents. Even if the link is not a directed link, there is a weak relation to communicate with the corresponding agent via other agents. In this paper, we define the number of links between agents as the 'distance'.

Under the environment, the Minor agent j_1 moves in the following. $P(Alt)$ is the probability for altruism behavior and its initial value is given.

(1) If the $P(Alt)$ is larger than random value at each step, the agent makes altruism behavior.
(2) The agent with altruism behavior finds the agents j_2 within distance r.
(3) All agents in the project i_1, which the agent j_1 participates, calculate Eq. (9) for the evaluation of project i_2 which the selected agent participates, and they decide to participate the project i_2 according to the evaluation.

$$R_{lm} \times \frac{P_i}{N_i} \geq A_{threshold}, \tag{9}$$

where R_{lm} is a binary value. If the distance between the agent in the project i_1 and the agent in the project i_2 is within r, R_{lm} is 1. Otherwise, 0. P_i is the task amount of project i_2, N_i is the number of agents in the project i_2, $A_{threshold}$ is a threshold value. If T_j is larger than T_{limit}, even if the agents with altruism behavior cannot participate to the project.

5. Simulation Results

This section describes the effects of the growing community by introducing the concept related to the tasks and the agent skills through the simulation results. In this paper, we prepared the simulation environment with 1,000 Major agents and 20,000 Minor agents. The threshold values $J_{threshold}$ in Eq. (6) and $L_{threshold}$ in Eq. (8) are used to find the optimal set by using the following patterns. The parameters are important to decide the action of the Minor agents.

$$\begin{aligned}(J_{threshold}, L_{threshold}) = &(0.0, 0.0), (0.5, 0.0), (1.0, 0.0),\\ &(0.0, 0.5), (0.5, 0.5), (1.0, 0.5),\\ &(0.0, 1.0), (0.5, 1.0), (1.0, 1.0)\end{aligned}$$

Figure 2 shows the relation of the number of developers and projects at the 1,000 iterations according to BA model. The horizontal axis is the number of developers and the vertical axis is the number of projects for the numbers of developers. For 9 kinds of parameter sets, the simulation results as shown in Fig. 2 were observed. The result is the average of 10 trials. The relation between the developers and the projects shows the linearity, if the number of developers is less than 1. Otherwise, the relation has the spread in the direction of the number of projects. In case of 1 or more developers, the curve of the relation depicts the spiral shape with polar coordinate. We tried to express the characteristic pattern as the following equations.

$$\begin{cases} r = a(\theta) \times \theta \\ a(\theta) = 0.07 - 0.07 \times \dfrac{\theta}{2\pi} \end{cases} \tag{10}$$

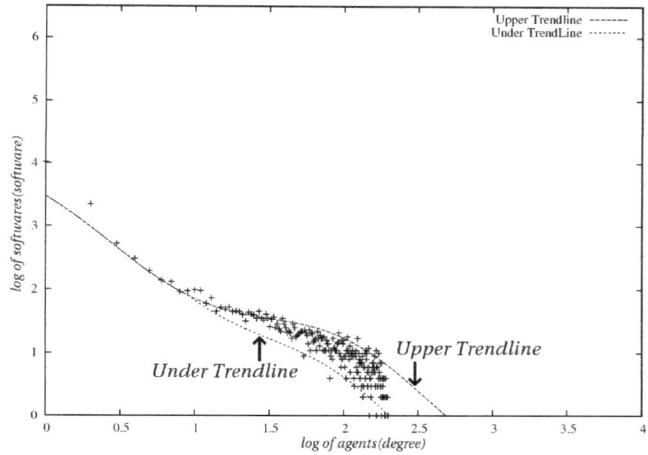

Fig. 2. The relation between developers and projects with BA model.

$$\begin{cases} r = a(\theta) \times \theta - \delta \ (\theta_1 \leq \theta) \\ \delta = 1.6 \times \dfrac{\theta - \theta_1}{2\pi}, \end{cases} \tag{11}$$

where δ means the width in the spread range. θ is the angle in the polar coordinate and θ_1 is the angle at the appropriate point where the number of developers is 1.

The characteristic pattern drawn by Eq. (10) and Eq. (11) can be seen in 9 kinds of parameter sets.

Moreover, we consider that the relation between the behaviors of the agent and the number of agents belonging to the project. Figure 3 shows the number of participating agents and the leaving agents for the tasks in the project. The horizontal axis is the number of tasks of the project and the vertical axis is the number of agents which take any actions. In case of the participating action and the leaving action, as the number of developers becomes large, the distribution approximates linearly. In addition, Fig. 4 shows the amount of tasks and the fitness when the task of the projects is created. We can see that the number of tasks in the project is large as the amount of tasks and the value of the fitness become high. Because the agent can select the tasks with high fitness easy by Eq. (4) and the agent can participate in the project with the large tasks easy by Eq. (5).

Figure 5 shows the changing number of leaving times as the amount of task increases. The increase tendency of the leaving agents appeared in the

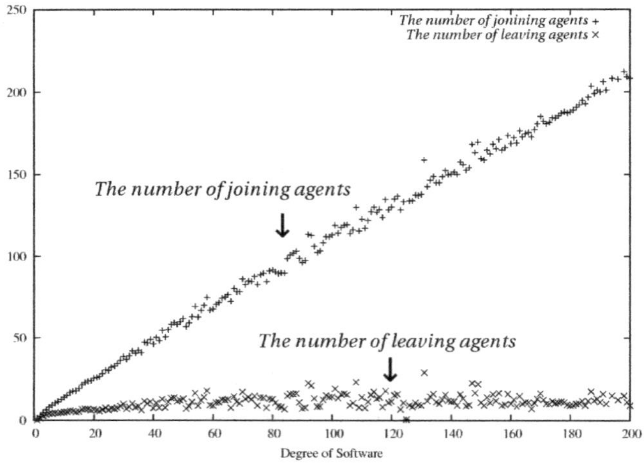

Fig. 3. The number of the participate agents and the leaving agents.

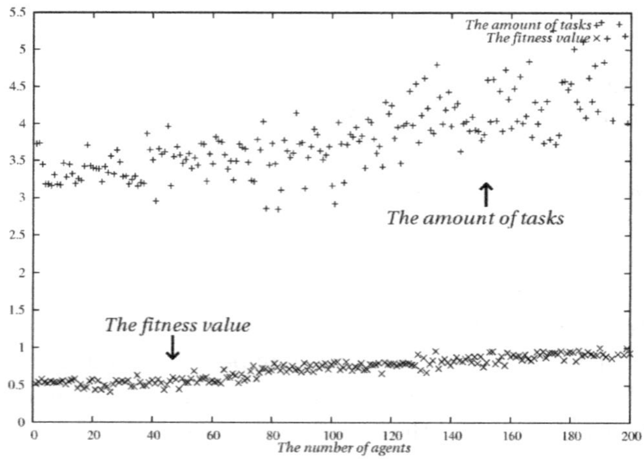

Fig. 4. The relation between the task and the fitness.

direction of large tasks of the software. Too many tasks in the developing software keep the agents at a distance. Especially, if there are many agents with the low capability in the developing software, that is, the load per agent is high, the leaving agents may increase. The situation may occur within 5 tasks in the software.

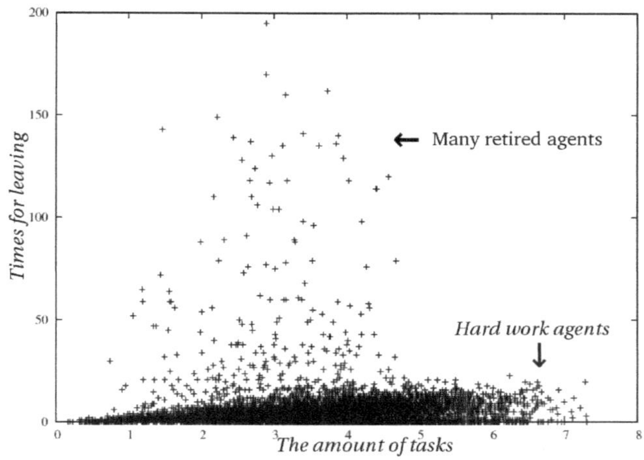

Fig. 5. The leaving agents for the amount of tasks.

6. Simulation results with altruism behaviors

Figure 6 shows the relation between the projects and developers with altruism behaviors. As compare with Fig. 2, the number of participated software was decreased in the log number of agents with $[0.0, 0.5]$. On the other hand, in the $[0.5, 1.5]$ the number was increased. The simulation result shows that the growing process occurs by the agents with altruism behaviors.

Table 1 shows the total average time for developing softwares. If there is an agent in the process for developing software, the software was counted. The developing for the agent with altruism behavior shows the good performance.

Figure 7 shows the leaving agents for the amount of tasks with altruism behaviors. As compare with Fig. 5, the agents with altruism behaviors shows that the leaving agents from the projects are decreased. These experimental results are the average of 10 trials.

Table 1. Total developing time.

	Software	Average time
without altruism	6,555	7.70
with altruism	26,119	3.74

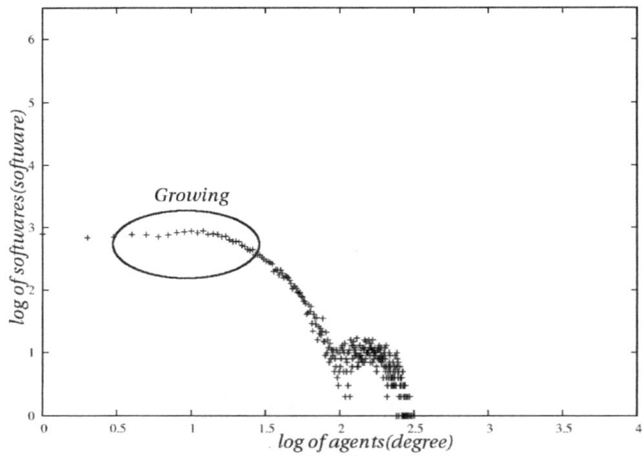

Fig. 6. The relation between the projects and developers with altruism behaviors.

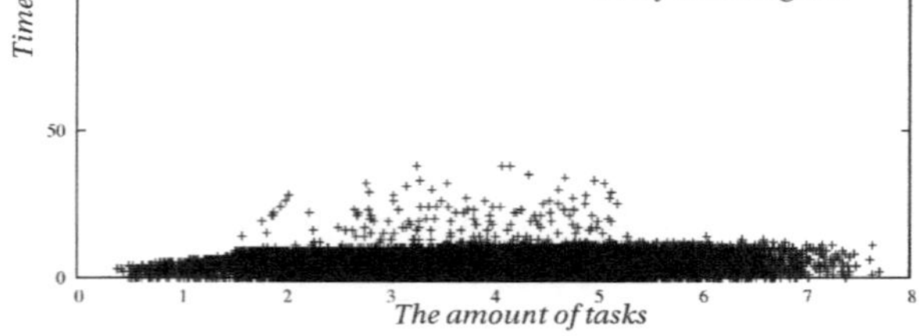

Fig. 7. The leaving agents for the amount of tasks with altruism behaviors.

7. Conclusion

With respect to community construction in software development, we examined with the agent based simulation with the relation between the task of the project and the skill of the agents. We can observe that the project in the software with the larger tasks and with the higher fitness are easy selected by many agents. The participating probability became greater in case of the large task and the larger fitness value. We can expect to realize the developing software smoothly by using our simulation model.

References

1. G. Madey, Y. Gao, V. Freeh, R. Tynan and C. Hoffman, *Agent-based Modeling and Simulation of Collaborative Networks*, Proc. of Americas Conference on Information System (AMCIS2002) pp. 1839-1842 (2003).
2. T. Ichimura and T. Uemoto, *Analysis of the Social Community Based on the Network Growing Model in Open Source Software Community*, Proc. of IEEE 8th Intl. Workshop on Computational Intelligence and Applications (IWCIA2015), pp. 149-153 (2015)
3. T. Ichimura, T. Uemoto and S. Kamada, *A Collaborative Group Developing Method of Open Source Software by Mimicking Altruism Behavior of Army Ant*, Proc. of the International MultiConference of Engineers and Computer Scientists 2016 (IMECS2016), Vol.1, pp. 46-51 (2016)
4. Source Forge.net, `https://sourceforge.net/` (retrieved 2016/10/31)
5. Source Forge Opendata, `http://srda.cse.nd.edu/mediawiki/index.php/Making_Queries` (retrieved 2016/10/31)
6. T. Ichimura, T. Uemoto and A. Hara, *Emergence of Altruism Behavior for Multi Feeding Areas in Army Ant Social Evolutionary System*, Proc. of 2014 IEEE International Conference on Systems, Man, and Cybernetics (IEEE SMC 2014), pp. 176-181 (2014).
7. T. Ichimura, T. Uemoto, A. Hara, and K. J. Mackin, *Emergence of Altruism Behavior in Army Ant Based Social Evolutionary System*, SpringerPlus, A Springer Open Journal, Vol. 3, 712, doi:10.1186/2193-1801-3-712(2014).
8. Y. Gao. G. Madey and V. Freeh, *Modeling and Simulation of the Open Source Software Community*, Agent-Directed Simulation Conference (2005).
9. A. L. Barabasi and R. Albert, *Emergence of scaling in random networks*, Science 286, pp. 509-512 (1999).
10. R. L. Trivers, *The evolution of reciprocal altruism*, Q Rev Biol, Vol. 46, pp. 35-67 (1971).

A REINFORCEMENT LEARNING WITH COLLECTIVE ACTIONS FOR GENERATING A MARSHALING PLAN OF FREIGHT TRAIN

Y. HIRASHIMA

Osaka Institute of Technology,
1-79-1 Kitayama, Hirakata, Osaka, 573-0196, Japan
yoichi.hirashima@oit.ac.jp

This chapter introduces a new reinforcement learning method to solve a train marshaling problem for assembling an outgoing train. In the problem, the arrangement of incoming freight cars are assumed to be random. Then, the cars are rearranged to the desired layout in order to assemble an outgoing train. In the proposed method, each set of freight cars that have the same destination make a group, and the desirable group layout constitutes the best outgoing train. Then, a rearrange operation is conducted by using several sub-tracks and an outgoing train is assembled in the main track. When a rearrangement operation is conducted in the proposed method, several cars located on different sub-tracks are collected by a locomotive. In order to rearrange cars by the desirable order, cars are removed from a sub-track to another sub-track. Each marshaling plan that consists of series of removal and rearrangement operations are generated by a reinforcement learning system based on the transfer distance of a locomotive. The total transfer distance of the locomotive required to assemble an outgoing train can be minimized by the proposed method.

Keywords: Collective action, Container transfer problem, Freight train, Marshaling, Q-Learning, Scheduling

1. Introduction

In recent years, railway transportation has an attention due to its small environmental load. However, since freight trains can transport goods only between railway stations, modal shifts are required for area that has no railway. Thus, the marshaling operation at freight station is required to joint several rail transports, or different modes of transportation including rail. A freight train is consists of several freight cars, and each car has its own destination. The train driven by a locomotive travels several destinations decoupling corresponding freight cars at each freight station. In intermodal transports including rail, containers carried into the station are loaded on freight cars and located at the freight yard in the arriving order. The initial layout of freight cars in the yard is determined by considering both arrangement of incoming train and the arriving order of the containers carried by non-rail methods. Containers carried into the station are loaded on freight cars and the initial layout of freight cars is thus random. For efficient shift in assembling outgoing train, freight cars must be rearranged before coupling to the freight train. In general, the rearrangement process is conducted in a freight yard that consists of a main-track and several sub-tracks. Freight cars initially placed on sub-tracks are rearranged, and lined into the main track. This series of operation is called marshaling, and several methods to solve the marshaling problem have been proposed [1-3]. However, in these methods, the marshaling is based on the layout generated by the classification process that assigns each car into a certain sub-track. This causes restrictions on the number of sub-tracks and the number of cars in a sub-track. On the other hands, in our research group, several methods that can generate marshaling plans from the random arrangement of cars in sub-tracks have been proposed [4, 5]. Since, in these methods, each rearrange operation are conducted for cars in the same sub-track, the marshaling plan can be improved by using method proposed in [6] considering collecting movements in a rearrange operation.

In this chapter the method for generating marshaling plan of freight cars in a train that has been proposed in [6] is introduced focusing on the collective movements in rearrangement operations. In the proposed method, the initial arrangement of freight cars in sub-tracks are assumed to be random, and thus any arrangements as the result of classification of incoming freight cars can be managed. Then, marshaling plans based on the transfer distance are obtained by a reinforcement learning method [7]. Freight cars to be rearranged are collected from several sub-tracks, coupled to each other and moved into the main track by a locomotive. Simultaneously, the optimal sequence of car-movements as well as the number of freight cars that can achieve the desired layout of outgoing train is obtained by autonomous learning. In order to show the effectiveness of the proposed method, computer simulations are conducted for several methods.

2. Problem description

2.1. *Freight Yard*

A freight yard is assumed to have 1 main track and m sub-tracks. Define k as the number of freight cars carried in and assigned to sub-tracks in a classification stage. Then, they are moved to the main track by the desirable order based on their destination in a marshaling stage. In the yard, a locomotive moves freight cars, and the movement of freight cars from sub-track to sub-track is called removal, and the car-movement from sub-track to main track is called rearrangement. For simplicity, the maximum number of freight cars that each sub-track can have is assumed to be n, the ith car is recognized by an unique symbol c_i, (i=1,…, k).

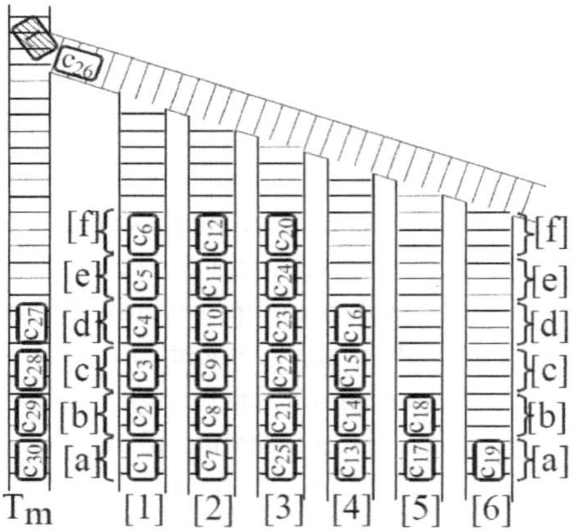

Fig. 1. Freight yard.

Figure 1 shows the outline of freight yard in the case k=30, m=n=6. In the figure, track Tm denotes the main track, and other tracks [1-6] are sub-tracks. The main track is linked with sub-tracks by a joint track, which is used for moving cars between sub-tracks, or for moving them from a sub-track to the main track.

When the locomotive L moves a certain car, other cars locating between the locomotive and the car to be moved must be removed to other sub-tracks. This operation is called removal. Then, if $k \leq nm-(n-1)$ is satisfied for keeping adequate space to conduct removal process, every car can be rearranged to the main track. In each sub-track, positions of cars are defined by n rows. Every position has unique position number represented by mn integers, and the position num ber for cars at the main track is 0.

	Position [1]	[2]	[3]	[4]	[5]	[6]	Marshaling [1]	[2]	[3]	[4]	[5]	[6]
[f]	31	32	33	34	35	36	c_6	c_{12}	c_{20}	c_{30}		
[e]	25	26	27	28	29	30	c_5	c_{11}	c_{24}	c_{27}	c_{28}	
[d]	19	20	21	22	23	24	c_4	c_{10}	c_{23}	c_{16}	c_{29}	
[c]	13	14	15	16	17	18	c_3	c_9	c_{22}	c_{15}	c_{26}	
[b]	7	8	9	10	11	12	c_2	c_8	c_{21}	c_{14}	c_{18}	
[a]	1	2	3	4	5	6	c_1	c_7	c_{25}	c_{13}	c_{17}	c_{19}

Fig. 2. Example of position index and yard arrangement.

Figure 2 shows an example of position index for $k = 30$, $m = n = 6$ and the layout of cars for Fig. 1-(b). In Fig. 2, the position "[a][1]" that is located at row "[a]" in the sub-track "[1]" has the position number 1. For unified representation of layout of cars in sub-tracks, the first car is placed at the row "[a]" in every track, and a newly placed car is coupled with the adjacent freight car.

2.2. *Desired layout in the main track*

In the main track, freight cars that have the same destination are placed at the neighboring positions. In this case, removal operations of these cars are not required at the destination regardless of arrangements of these cars. In order to consider this feature in the desired layout in the main track, a group is organized by cars that have the same destination, and these cars can be placed at any positions in the group. Then, making a group corresponding to each destination, the order of groups lined in the main track is predetermined by destinations. This feature yields several desirable layouts in the main track.

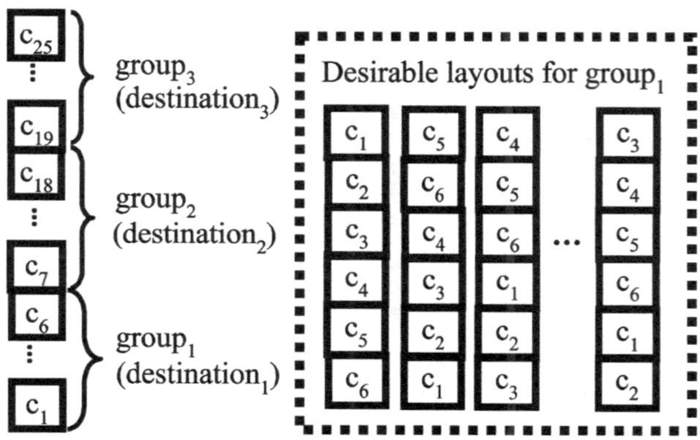

Fig. 3. Groups and arrangements of freight cars.

Figure 3 depicts examples of desirable layouts of cars and the desired layout of groups in the main track. In the figure, freight cars $c_1,\ldots,c_6$ to the destination1 make group1, $c_7,\ldots,c_{18}$ to the destination2 make group2, $c_{19},\ldots,c_{25}$ to the destination3 make group3. Groups1,2,3 are lined by ascending order in the main track, which make a desirable layout. Also, in the figure, examples of layout in group1 are in the dashed square.

The layout of groups lined by the reverse order does not yield additional removal actions at the destination of each group. Thus, in the proposed method, the layout lined groups by the reverse order and the layout lined by ascending order from both ends of the train are regarded as desired layouts. By defining r as the number of groups, the total number of layouts of group is 2^{r-1}.

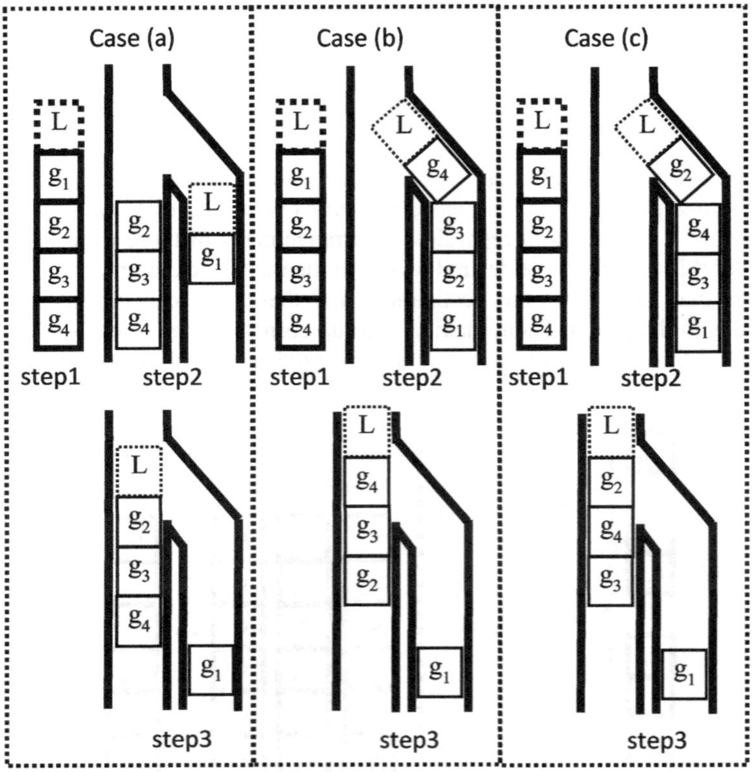

Fig. 4. Group layouts in a out going train.

Figure 4 depicts examples of material handling operation for extended layout of groups at the destination of group1. In the figure, step 1 shows the layout of the incoming train. In case (a), cars in group1 are separated at the main track, and moved to a sub-track by the locomotive L at step 2. In cases (b),(c), cars in group1 are carried in a sub-track, and group1 is separated at the sub-track. In the cases, group1 can be located without any removal actions for cars in each group. Thus, these layouts of groups are regarded as candidate for desired one in the learning process of the proposed method.

When there exists a rearranging car that has no car to be removed on it, its rearrangement precedes any removals. In the case that several cars can be rearranged without a removal, rearrangements are repeated until all the candidates for rearrangement requires at least one removal. If several candidates for rearrangement require no removal, the candidates are collected to the sub-track that has one of the candidates, and an order of collections into the sub-track is selected from several combinations. In this case, the arrangement of cars in sub-tracks obtained after rearrangements is unique, so that the removal operations have no correlation with the order of collective movements. This operation is called direct rearrangement. When a car in a certain sub-track can be rearrange directly to the main track and when several cars located adjacent positions in the same sub-track satisfy the layout of group in the main track, they are coupled and applied direct rearrangement.

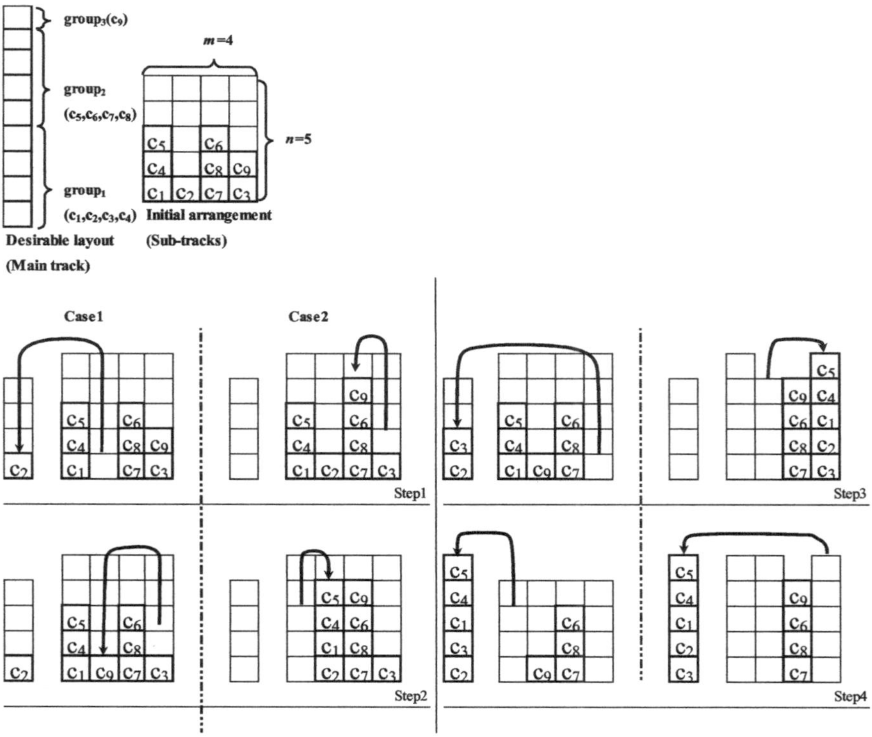

Fig. 5. Example of marshaling.

Figure 5 shows 2 cases of marshaling including a removal and rearrangements, existing candidates for rearranging cars that require no removal. At the top of figure, from the left side, a desired layout of cars and groups, and the initial layout of cars in sub-tracks followed by 2 cases in 2 columns are depicted for $m = 4$, $n = 5$, $k = 9$. In both cases, c_1, c_2, c_3, c_4 are in group1, c_5, c_6, c_7, c_8 are in group2, and group1 must be rearranged first to the main track. In each group, any layouts of cars can be acceptable. In case1, c_2 in step 1, c_3 in step 3 and step 4 are applied the direct rearrangement. Also, in case2, steps 2 through 4 are direct rearrangements. Step2 in case1 and step1 in case2 are removals. Step 4 in case1, 3 cars c_1, c_4, c_5 located adjacent positions are coupled with each other and moved to the main track by a direct rearrangement operation. In addition, at steps 2 and 3 in case 2, cars in group1 and group2 are collected in sub-tracks to rearrange directly, since the arrangement of c_3, c_2, c_1, c_4, c_5 can satisfy the desired layout of groups in the main track. Whereas, at steps 1,3,4 in case 1, 3 direct rearrangements are conducted separately.

2.3. *Marshaling process*

A marshaling process consists of following 6 operations:

(I) selections: group-layout in the main track,

(II) rearrangement of freight cars to the main track, when they can be moved directly,

(III) selection: a freight car to be rearranged into the main track,

(IV) selection: removal destinations of the cars in front of the car selected in (IV),

(V) selection: the number of cars to be moved in (V),

(VI) removal of the cars determined in (VI) to the selected sub-track in (V).

After operations (I),(II) are finished, (III)-(VII) are repeated until one of desirable layouts is achieved in the main track, and a series of operations from the initial one in (I) to the final one in (III) achieving the desirable layout is defined as a trial.

Now, define Go as the desired layout, and h_1 as the number of candidates of Go. Each candidate in operation (I) is represented by u_{j1} ($1 \leq j_1 \leq h_1$, $h_1 = 2^{r-1}$.)

In the operation (II), an order to collect cars located several sub-tracks is defined as T_C, and h_2 candidates of T_C are defined as u_{j2} ($h_1+1 \leq j_2 \leq h_1+h_2$). Each u_{j2} constructs a sequence of cars which satisfies a part of the layout in Go.

In the operation (IV), each group has the predetermined position in the main track. Then, the car to be rearranged is defined as c_T, and candidates of c_T is determined by the number of freight cars that have already rearranged to the main track and the group layout in Go. h_3 is defined as the number of freight cars in group to be rearranged, and u_{j3} ($h_1+h_2+1 \leq j_3 \leq h_1+h_2+h_3$) as candidates of c_T.

In the operation (V), the removal destination of cars located on the car to be rearranged is defined as T_R. Then, defining u_{j4} ($h_1+h_2+h_3+1 \leq j_4 \leq h_1+h_2+h_3+m-1$) as candidates of T_R, excluding the sub-track that has the car to be removed, and the number of candidates is m-1.

In the operation (VI), defining n_p as the number of removal cars required to rearrange c_T, and defining n_q as the number of removal cars that can be located the sub-track selected in the operation (V), the candidate numbers of cars to be moved are determined by u_{j6} ($1 \leq u_{j5} \leq \min\{n_p, n_q\}$, $h_1+h_2+h_3+m \leq j_5 \leq h_1+h_2+h_3+m + \min\{n_p, n_q\}$). (III)-(VI) are repeated until all the cars are rearranged into the main track.

2.4. *Transfer distance of Locomotive*

When a locomotive transfers freight cars, the process of the unit transition is as follows: (E1) starts without freight cars, and reaches to the joint track, (E2) restarts in reverse direction to the target car to be moved, (E3) joints them, (E4) pulls out them to the joint track, (E5) restarts in reverse direction, and transfers them to the indicated location, and (E6) disjoints them from the locomotive. Then, the transfer distance of locomotive in (E1), (E2), (E4) and (E5) is defined as D_1, D_2, D_3 and D_4, respectively, and the distance of the unit transition D is calculated by $D = D_1 + D_2 + D_3 + D_4$. Also, define the unit distance of a movement for cars in each sub-track as D_{minv}, the length of joint track between adjacent sub-tracks, or, sub-track and main track as D_{minh}. The location of the locomotive at the end of above process is the start location of the next movement process of the selected car. The initial position of the locomotive is located on the joint track nearest to the main track.

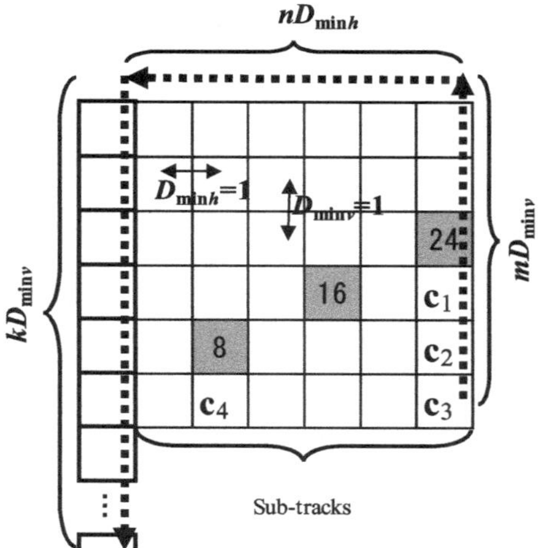

(b) movement of cars

31	32	33	34	35	36
25	26	27	28	29	30
19	20	21	22	23	24
13	14	15	16	17	18
7	8	9	10	11	12
1	2	3	4	5	6

(a) Position index in sub-tracks

Fig. 6. Transfer distance.

Figure 6 shows an example of transfer distance. In the figure, $m = n = 6$, $D_{\min v} = D_{\min h} = 1$, $k = 18$, (a) is position index, and (b) depicts movements of locomo-tive and freight car. Also, the locomotive starts from position 8, the target is located on the position 18, the destination of the target is 4, and the number of cars to be moved is 2. Since the locomotive moves without freight cars from 8 to 24, the transfer distance is D1 + D2 = 12, D1 = 5, D2 = 7, whereas it moves from 24 to 16 with 2 freight cars, and the transfer distance is D3 + D4 = 13 (D3 = 7, D4 = 6).

3. Learning algorithm

Defining Q_1 as an evaluation value for *Go*, $Q_1(Go)$ is updated by the following rule when one of desired layout is achieved in the main track:

$$Q_1(G_O) \leftarrow \max\left\{Q_1(G_O), (1-\alpha)Q_1(G_O) + \alpha R\prod_{i=1}^{l}\gamma_i\right\} \tag{1}$$

where l denotes the total movement counts required to achieve the desired layout, α is learning rate, γ_i is discount factor calculated for each movement, R is reward that is given only when one of desired layout is achieved in the main track.

Define $\boldsymbol{s}(t)$ as the state at time t, Tc as the sub-track selected as the destination for the removed car, p_C as the number of classified groups, q_M as the movement counts of freight cars by direct rearrangement, and $\boldsymbol{s}'$ as the state that follows $\boldsymbol{s}$. In the direct rearrangement, Q_2 is defined as evaluation values for $(\boldsymbol{s}_1, u_{j2})$, where $\boldsymbol{s}_1 = [\boldsymbol{s}, Go]$, $\boldsymbol{s}_2 = [\boldsymbol{s}_1, T_C]$. $Q_2(\boldsymbol{s}_1, T_C)$ is updated by the following rule:

$$Q_2(\boldsymbol{s}_1, T_C) \leftarrow (1-\alpha)Q_3(\boldsymbol{s}_3, c_T) + \alpha V_1$$
$$V_1 = \begin{cases} R\prod_{i=1}^{q}\gamma_i & \text{(all cars moved into the main track)} \\ \gamma \max_{u_{j3}} Q_3(\boldsymbol{s}_3, u_{j_3}) & \text{(otherwise)} \end{cases} \tag{2}$$

In the rearrangement, define q as the number of direct movements conducted sequentially, p_M as the number of cars moved. Q_3,Q_4 and Q_5 are defined as evaluation values for $(\boldsymbol{s}_1,u_{j3})$,$(\boldsymbol{s}_3,u_{j4})$,$(\boldsymbol{s}_4,u_{j5})$ respectively, where $\boldsymbol{s}_3=\boldsymbol{s}_1$, $\boldsymbol{s}_4=[\boldsymbol{s}_3, c_T]$, $\boldsymbol{s}_5=[\boldsymbol{s}_4, T_R]$. $Q_3(\boldsymbol{s}_3,c_T)$, and $Q_4(\boldsymbol{s}_4, T_R)$ $Q_5(\boldsymbol{s}_5,p_M)$ are updated by following rules:

$$Q_3(\boldsymbol{s}_3, c_T) \leftarrow \max_{u_{j4}} Q_4(\boldsymbol{s}_4, u_{j4}) \tag{3}$$

$$Q_4(\boldsymbol{s}_4, T_R) \leftarrow \max_{u_{j5}} Q_5(\boldsymbol{s}_5, u_{j5}) \tag{4}$$

$$Q_5(\boldsymbol{s}_5, p_M) \leftarrow \begin{cases} (1-\alpha)Q_5(\boldsymbol{s}_5, p_M) + \alpha\left[V_2\prod_{i=1}^{q+1}\gamma_i\right], \\ V_2 = \max_{u_{j3}} Q_3(\boldsymbol{s}_3', u_{j3}) \quad (u \text{ is a rearrangement}) \\ (1\text{-}\alpha)Q_5(\boldsymbol{s}_5, p_M) + \alpha\left[R + \gamma V_3\right], \\ V_3 = \max_{u_{j4}} Q_4(\boldsymbol{s}_4', u_{j4}) \quad (u \text{ is a removal}) \end{cases} \tag{5}$$

where α is the learning rate, R is the reward that is given when one of desirable layout is achieved, and γ is the discount factor that is used to reflect the transfer distance of locomotive and calculated by the following equation:

$$\gamma = \delta \frac{D_{max} - \beta D}{D_{max}}, \qquad (0 < \beta < 1, 0 < \delta < 1) \tag{6}$$

where D_{max} is the maximum value of D.

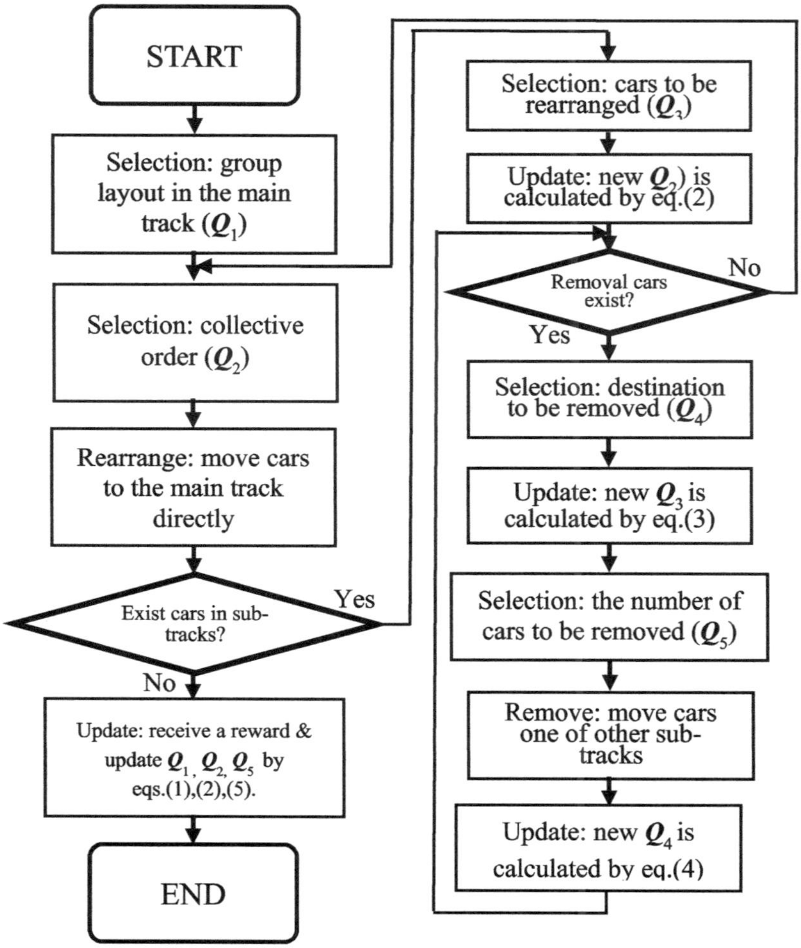

Fig. 7. Flowchart of the learning algorithm.f.

A flowchart of the learning algorithm for 1 trial is depicted in Fig. 7. Propagating Q-values by using eqs. (1)-(6), Q-values are discounted according to the transfer distance of locomotive. In other words, by selecting the removal destination that has the largest Q-value, the transfer distance of locomotive can be reduced.

In the learning stages, each u_j, $(1 \leq j \leq h_1+h_2+h_3+m+\min\{n_p, n_q\})$ is selected by the soft-max action selection method [6]. In the addressed problem, Q_2, Q_3, Q_4, Q_5 become smaller when the number of discounts becomes larger. Then, for complex problems, the difference between probabilities in candidate selection remain small at the initial state and large at final state before achieving desired layout, even after repetitive learning. In this case, obtained evaluation does not contribute to selections in initial stage of marshaling process, and search movements to reduce the transfer distance of locomotive is spoiled in final stage. To conquer this drawback, Q_2, Q_3, Q_4, Q_5, Q_6 are normalized, and probability P for selection of each candidate is calculated by

$$\tilde{Q}_i(s_i,u_{j_i}) = \frac{Q_i(s_i,u_{j_i}) - \min_{u_{ji}} Q_i(s_i,u_{j_i})}{\max_{u_{ji}} Q_i(s_i,u_{j_i}) - \min_{u_{ji}} Q_i(s_i,u_{j_i})}, \tag{7}$$

$$P(s_i,u_{j_i}) = \frac{\exp(\tilde{Q}_i(s_i,u_{j_i})/\xi)}{\sum_{u\in u_{ji}} \exp(\tilde{Q}_i(s_i,u)/\xi)} \quad (\mathrm{i}=2,3,4,5), \tag{8}$$

$$P(u_{j_1}) = \frac{\exp(\tilde{Q}_1(u_{j_1})/\xi)}{\sum_{u\in u_{j1}} \exp(\tilde{Q}_1(u)/\xi)} \quad (\mathrm{i}=1), \tag{9}$$

wherer ξ is a thermo constant.

4. Computer simulations

Computer simulations are conducted for $m = 12$, $n = 6$, $k = 36$ and learning performances of following 2 methods are compared:

(A) Proposed method considering the layout of groups with collective movements in the direct rearrangement,

(B) Proposed method without collective movements in the direct rearrangement [5].

The initial arrangement of freight cars in sub-tracks train is described in Fig. 8.

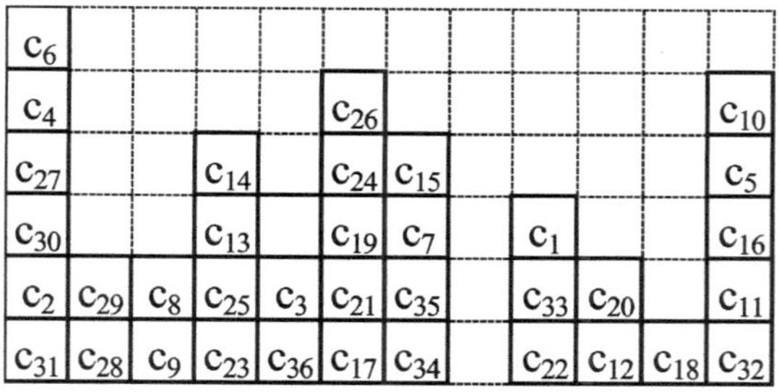

Fig. 8. Initial arrangement of cars in sub-tracks.

The original group-layout is $group_1$, $group_2$, $group_3$, $group_4$: the order as shown in Fig. 9. Also, cars $c_1,\ldots,c_9$ are in $group_1$, $c_{10},\ldots,c_{18}$ are in $group_2$, $c_{19},\ldots,c_{27}$ are in $group_3$, $c_{28},\ldots,c_{36}$ are in $group_4$. Other parameters are set as $\alpha=0.9$, $\beta=0.2$, $\delta=0.9$, R=1.0, $\xi=0.1$. Both methods consider extended layout of groups and derive desirable layouts autonomously in order to reduce the total transfer distance of the locomotive. The locomotive and freight cars assumed to have the same length, and $D_{\min v}=D_{\min h}=20$m.

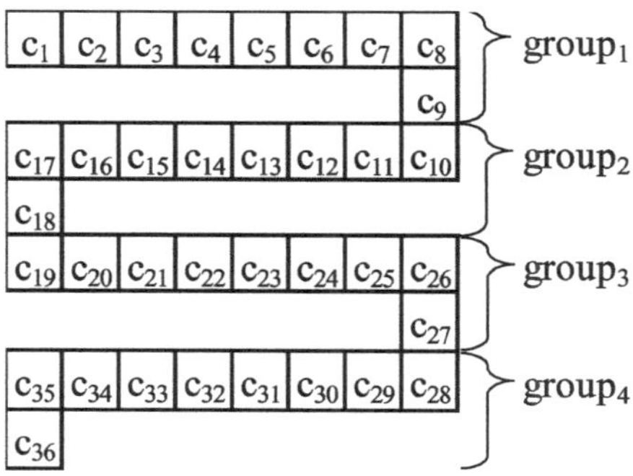

Fig. 9. Original group-layout for an outgoing train in the main track.

The results are shown in Fig. 9. In the figure, horizontal axis expresses the number of trials and the vertical axis expresses the minimum processing time found in the past trials. A trail starts with an initial arrangement in sub-tracks, and finished when all the freight cars are rearranged into the main track. Each result is averaged over 20 independent simulations. In Fig. 9, the learning performance of method (A) is better than that of (B), because (A) collects freight cars in sub-tracks for each direct rearrangement.

Table 1 Total transfer distance.

	Transfer distance [m]		
	best	average	worst
Method(A)	15280	15480	15800
Method(B)	24980	25600	26720

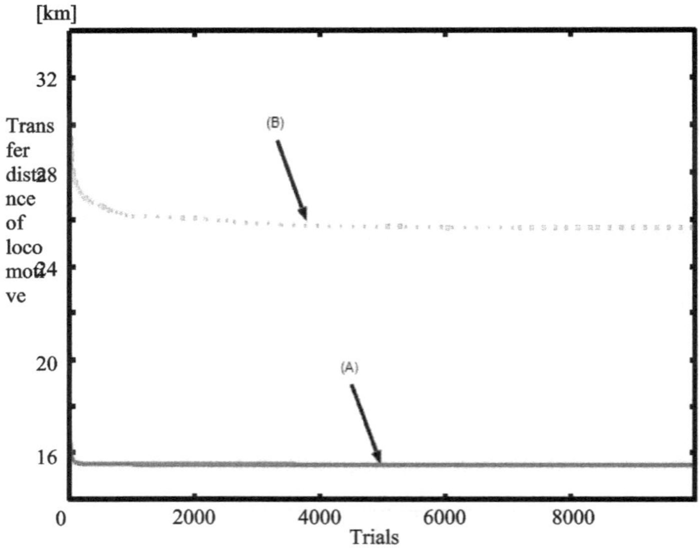

Fig. 9. Performance comparison.

5. Conclusions

A new scheduling method has been proposed in order to rearrange and line cars in the desirable order onto the main track considering collective movements of the locomotive in the direct rearrangement of freight cars.

The learning algorithm of the proposed method is derived based on the reinforcement learning that evaluates the total transfer distance of locomotive in a marshaling. In order to reduce the total transfer distance of locomotive in a marshaling, the proposed method obtain group-layout of desirable arrangement of outgoing train and collecting order of freight cars in sub-tracks, so that the total transfer distance in a marshaling plan derived by the proposed method has been reduced by about 40% as compared to the conventional method in computer simulations. In addition, the arrangement of freight cars in the main track, the rearrange order of cars, the position of each removal car, the number of cars to be removed, and the group layout in the outgoing train has been obtained simultaneously so that the learning performance of the proposed method has been improved.

References

1. C. F. Daganzo, R. G. Dowling and R. W. Hall: Railroad classification yard throughput: The case of multistage triangular sorting, Transportation Research, Part A, No. 17, Vol. 2 (1983), pp. 95-106.
2. U. Blasum, M. R. Bussieck, W. Hochstättler, C. Moll, H.-H. Scheel and T. Winter: Scheduling Trams in the Morning, Mathematical Methods of Operations Research, Vol. 49, No. 1 (2000) , pp. 137-148.
3. R. Jacob, P. Marton, J. Maue and M. Nunkesser: *Multistage Methods for Freight Train Classification*, Proceedings of 7th Workshop on Algorithmic Approaches for Transportation Modeling, Optimization, and Systems (2007), pp. 158-174.
4. Y. Hirashima: *An Intelligent Train Marshaling Based on the Processing Time Considering Group Layout of Freight Cars* in IAENG Transactions on Electrical Engineering, Vol. 1, World Scientific (2013), pp. 229-243.
5. Y. Hirashima: A Reinforcement Learning System for Generating Train Marshaling Plan of Freight Cars Based on the Processing Time Considering Group Layout, Lecture Notes in Engineering and Computer Science: Proceedings of The International MultiConference of Engineers and Computer Scientists 2016, IMECS 2016, 16-18 March, 2016, Hong Kong, pp. 30-35.
6. Y. Hirashima: A Reinforcement Learning for Marshaling of Freight Train Considering Collective Motions, Lecture Notes in Engineering and Computer Science: Proceedings of The International MultiConference of Engineers and Computer Scientists 2016, IMECS 2016, 16-18 March, 2016, Hong Kong, pp. 19-24.
7. R. Sutton and A. Barto: *Reinforcement Learning*, MIT Press (1999).

LANDSLIDE HAZARD ZONATION USING UNSUPERVISED ARTIFICIAL NEURAL NETWORKS

M.A. DELA CERNA
Surigao State College of Technology,
Surigao City, 8400, Philippines
madc712004@yahoo.com

The study evaluates spatial data mining through spatial clustering. Unsupervised learning algorithm was used such that of K-Means and Self-Organizing Map in Artificial Neural Networks. It maps landslide hazard areas utilizing eight causative landslide indicators as datasets such as slope gradient, vertical displacement, drainage density, rate of weathering, lithology, ground stability, soil type, and vegetation. Techniques limitations concentrate only on a particular kind of computation. The findings of the study showed vertical displacement and slope gradient as two major aspects of factors that dominate which cause landslide. Landslide hazard zones are identified with a model that map areas under study and a tool for visualizing identified causative factors of landslides.

Keywords: artificial neural networks; k-means; landslide hazard zonation; self-organizing map; unsupervised learning algorithm.

1. Introduction

One of the major natural disasters in the world is landslide which causes economic damage and loss of life. A landslide hazard investigation is imperative and the laws pertaining thereto must be unequivocally implemented. Natural disasters like landslide can be prevented and made possible through landslide hazard zonation which emphasizes the importance of assessment (qualitative) and evaluation (quantitative) of hazards from landslides [1, 2].

Researchers who did studies on landslides have applied models for predicting the same. Some of them used probabilistic models. Various models utilized were logistic regression [3], neuro-fuzzy [4], fuzzy logic [5], geographic information system [5, 6], self-organizing map [7] and artificial neural network models [3]. Artificial neural networks (ANN) have broad applicability to real-world business problems. Successful application of it is observed in many industries [8]. Since neural networks are best for identifying patterns or trends in data, they are well suited for prediction or forecasting including sales forecasting, industrial process control, customer research, data validation, risk management [3, 9] and target marketing [10]. Kohonen self-organizing map (SOM) which is a part of the unsupervised learning algorithm of ANN is applied clustering algorithm of high dimensional data, also as an alternative tool to classical multivariate statistical techniques [11]. Moreover, SOM algorithm is also a tool for investigation of eco-morphological life history of fish [12], also model for post-fire hydrologic and geomorphic hazards [13], as well as instrument for forecasting reservoir inflow during typhoon [9], and lastly for visualizing topical structure of medical sciences [14]. On the other hand, K-means clustering algorithm is for image segmentation based on adaptive approach [15], landslide detection [16] and spatial prediction for landslide hazard [17]. Evidences of above lists of references, utilizing SOM and K-Means have been recently applied to a broad range of geo-environmental fields.

With technology, it aids some studies to invent software that maps landslide risk associated caused by improper environmental practices such as that of SSM activities. They are less likely monitored since there is lack of existing profiling and mapping of occurrences. Continued investigation into occurrences and implications of geo-hazards of mining explorations led to specific mitigation studies and establishing of preventive measures.

Thus, this study aims to identify landslide prone areas by applying clustering techniques and assess the geo-environmental factors using K-Means and Self-organizing Map algorithms in preparation for landslide hazard zonation. It is used to achieve landslide assessment in small scale mining areas. It is a significant step towards landslide hazard and risk management. The method will aid the government in its decision making relating to disaster risk management and control.

2. Methods

The study employed the unsupervised learning algorithm in artificial neural networks using K-Means and Self-organizing Map algorithms. The datasets used to test the algorithm were primary data from comprehensive survey conducted. Data were imported from comma delimited (.csv) files and interpreted by the K-Means and SOM for clustering.

2.1. *Location of the Study*

The Philippines is known to be a landslide prone country, after Indonesia, India, and China [18]. The country is hilly and mountainous that experience earthquakes as well as intense rainfall brought about by monsoons or typhoons. The terrain is naturally susceptible to landslides [18]. The country needs continuous monitoring in high-risk areas and should have comprehensive study on causes and effects of landslide. Since mining is booming in the country, small scale mining (SSM) activities exist in some areas of Surigao del Norte province.

Surigao del Norte is a province of the Philippines located in the Caraga region of Mindanao between 125°15' to 126°15' east longitude and 9°18' to 10°30' north latitude [18]. It is bounded on the north and east by the Pacific Ocean, on the south by the provinces of Agusan del Norte and Surigao del Sur, and on the west by the Surigao Strait. The province has abundant mineral reserves including gold, iron, manganese, silica, cobalt, copper, chromite, limestone, silver and has the world's largest nickel deposits [19]. The province has a tropical climate type and prone to brief afternoon downpour and thunderstorms [19]. Most of the active mining areas are located in moderate to steep rugged terrain underlain by volcanic rocks and terrace gravel deposits and structurally controlled along Philippine fault line. The practices employed by small scale miners involve deep tunneling such as sinking and drifting; trenching and hydraulic mining in the mountains. These are documented. The inherent digging of soil and sluicing involved therein cause siltation, erosion, and land degradation which certainly are issues of environmental concerns [20]. There also have been unreported incidents of fatalities due to landslide caused by unregulated SSM activities [21]. Miners and their families who live at the peak of the mountains in mining areas are directly affected by this. Figures 1 and 2 present the map of Surigao del Norte and the panoramic views of some identified landslide prone mining areas respectively.

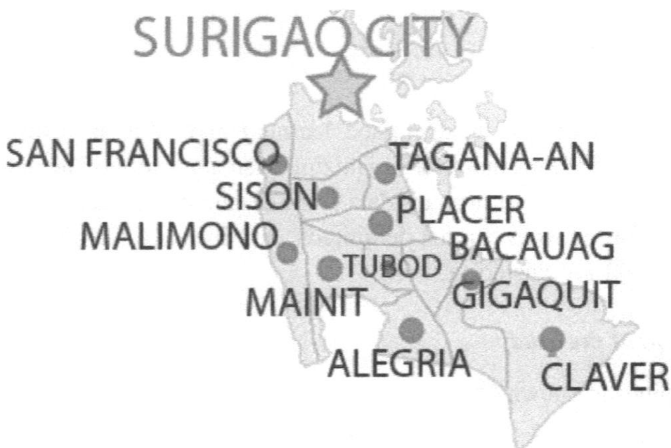

Fig. 1. Map of Surigao del Norte, Philippines.

Fig. 2. Panoramic view of some mining areas (a) GoldDustSlide1, Magsaysay, Placer; (b) BrazilSlide2, Sitio Brazil, Mat-i, Surigao City, (c) CansayongSlide2, Brgy. Cansayong, Malimono; (d) SanPedroSlide2, Brgy. San Pedro, Alegria.

2.2. *Data Preparation*

Primary data were from comprehensive survey in mining areas. There were eight causative factors considered. Indicators of landslide hazard are: slope gradient which classes are flat and undulating (FU), moderately sloping (MS), hilly and moderately steep (HS), and steep (SP).

Vertical displacement classifications are shallow (SH), moderate (MV), deep (DP), and very deep (VD). On the other hand, drainage density has low (LD) moderate (MD), high (HD), and very high (DD) classes. While rate of weathering are from slightly weathered (SW), moderately weathered (MW), highly weathered (HW), to completely weathered (CW).

Lithology in the areas are andesite porphyry, diorite, agglomerate, hydrothermal breccia, sandstone, siltstone, alluvium, terrace gravel and ultramafic which tend to be weakly fractured/weathered (WF), moderately fractured/weathered (MF), highly weathered/fractured and delighting structure (HF), and highly fractured and highly weathered with tension cracks/or landslide scarp (VF).

Ground stability classes are stable with no identified landslide scars, either old, recent or active (ST), presence of soil creep and other indications for possible landslide occurrence (SC), inactive landslides evident; tension cracks present (IL), and active landslides are evident with presence of tension cracks, bulges, terracettes, seepage (AL).

For soil type, classes are silts and slays (50% or more of material is smaller than No. 200 sieve size) with liquid limit of 50% or greater (SG), silts and clays (50% or more of material are lower than No. 200 sieve size) with liquid limit of less than 50% (SL), sands (SD), and gravels (GV) based on the Unified Soil Classification System [22]. In vegetation, the classes are primary growth (PG), secondary growth (SG), and tertiary growth (TG).

The rate of the slope gradient, vertical displacement, and drainage density uses the actual quantitative rate while the rate of weathering, lithology, ground stability and soil type is in qualitative type indicators, their classes use an ordinal number (1,3,6,9) likewise in vegetation (1,5,9) which has only three classes. Table 1 presents the rate of landslide hazard parameters.

Table 1. Rate of landslide hazard parameters.

Indicators	Classes	Rate	Indicators	Classes	Rate
Slope Gradient	FU	0° - < 8°	Lithology	WF	1
	MS	8° - < 18°		MF	3
	HS	18° -< 26°		HF	6
	SP	26° -<= 90°		VF	9
Vertical Displacement	SH	<2 m	Ground Stability	ST	1
	MV	2-9 m		SC	3
	DP	10-25 m		IL	6
	VD	>25 m		AL	9
Drainage Density	LD	0-1 drainage	Soil Type	SG	1
	MD	2-4 drainages		SL	3
	HD	5-10 drainages		SD	6
	DD	>10 drainages		GV	9
Rate of Weathering	SW	1	Vegetation	PG	1
	MW	3		SG	5
	HW	6		TG	9
	CW	9			

2.3. *Application of K-Means Clustering Algorithm*

The study used the multivariate statistical analysis of interaction of factors. Landslide hazard zonation compares each data layer of causative factor to existing landslide distribution. Weights of landslide causative factors are assigned based on landslide density. K-Means clustering is a significant statistical method used in landslide hazard zoning.

K-means algorithm commences by placing K points at random locations in space. Then performs the following steps iteratively: (1) assign clusters with nearest centroid, and (2) move each centroid to the mean of the instances assigned to it. The algorithm continues until no instances change cluster membership.

This algorithm attempts to minimize intra-cluster variance. The algorithm converges to a local minimum; therefore different initializations will result in different clustering. K-means does not guarantee similar instances end up in the same cluster. It is a method of vector quantization, initially patterned from signal processing that is famous for cluster analysis in data mining. K-Means aims to partition *n* observations into *k* clusters in which each observation fit into the cluster with the nearest mean; that serves as a prototype of the group. Below are the steps of using the K-Means algorithm.

1. Initialize the K weight vectors, e.g. to randomly chosen examples. Each weight vector represents a cluster.

2. Assign each input example x to the cluster $c(x)$ with the nearest corresponding weight vector refer to Eq. (1).

$$c(x) = \arg\min_j \left\| x - w_j(n) \right\| \tag{1}$$

3. Update the weights: See Eq. (2).

$$w_j(n+1) = \sum_{x \text{ such that } c(x)=j} x / n_j \tag{2}$$

with n_j the number of examples assigned to cluster j.

4. Increment n by 1 and go until no noticeable changes of weight vectors occur.

2.4. *Application of Self-Organizing Map Algorithm*

Multivariate statistical analysis for landslide hazard zonation considers relative contribution of each thematic data layer to the total landslide susceptibility and use classification or clustering of hazard for the given area. Self-Organizing Map is the method used for landslide hazard zoning.

Teuvo Kohonen introduced self-organizing map in 1982 also known as Kohonen feature map. SOM is under unsupervised neural networks for clustering tool of high-dimensional and complex data. The SOM provides high-dimensional input space of clustering data in most cases two-dimensional, output space. This output space is in a rectangular grid of units, a representation that is easily understandable for its users due to its analogy 2-D maps. Each of the units on map is assigned a weight vector, which is of the same dimensionality as the vectors in the input space [23].

The universal application of SOM of Kohonen is for mapping *n*-dimensional input vectors to two-dimensional *neurons* or *maps*. This map describes variations in the information of the input data, and the topology of the original data is preserved on SOM through connection of the input vectors which shares common features of similar or neighboring neurons. This is motivated by feature of the human brain wherein neurons are organized in one or multi-dimensional lattice. The neurons compete among themselves to be activated according to a competitive learning scheme. The weight vector associated with winning neuron is only updated in the scheme winner-takes-all.

Typically, SOM algorithm performs some successive iterations until reference vectors associated with neuron of a two-dimensional map represent, as much as possible, the input patterns that are closer to those neurons. The aim is to learn a feature map from the spatially continuous input space, in which the input vectors live, to low dimensional spatially discrete output space, which is formed by arranging the computational neurons into a grid. Figure 3 illustrates the concept of SOM.

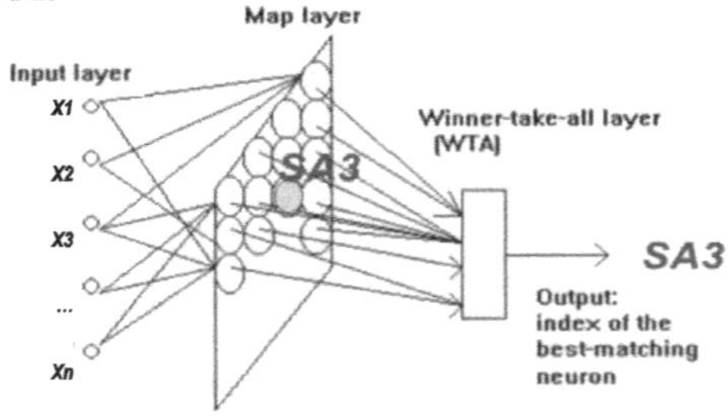

Fig. 3. Block diagram of Self-Organizing Map.

The stages of SOM algorithm that achieves this can be summarized as follows:

(1) Initialization. Choose random values for the initial weight vectors w_j.

(2) Sampling. Draw a sample training input vector x from the input space.

(3) Matching. Find the winning neuron that has weight vector closest to the input vector. To determine the best matching unit, one method is to iterate through all the nodes and calculate the Euclidean distance between each node's weight vector and the current input vector. The node with a weight vector closest to the input vector is tagged as the best matching unit. See Eq. (3), the Euclidean distance is given as:

$$d_{euc}\left(x,w\right) = \sqrt{\sum_{i=1}^{n} \left(x_i - w_i\right)2} \tag{3}$$

where $\boldsymbol{x}$ is the current input vector and $\boldsymbol{w}$ is the node's weight vector.

(4) Updating. Apply the weight update equation. See Eq. (4).

$$w_j(n+1) = w_j(n) + \eta(n)\ h_{i(x)}(d_{i(x)j})\ (x(n)\text{-}w_j(n)) \tag{4}$$

where $h_{i(x)}$ is a Gaussian neighborhood and is the learning rate η.

(5) Continuation. Keep returning to step 2 until the feature map stops changing.

2.5. *System Architecture*

Figure 4 illustrates architectural design which has two functions; landslide hazard assessment and landslide hazard zonation. The datasets for the eight causative factors were used to input into the system and it were simulated using two types of unsupervised learning algorithm, the K-Means and SOM. A sort of testing for the system was done to provide a robust and accurate result after coming up a decision what type of landslide hazard rate to assert for each mining areas.

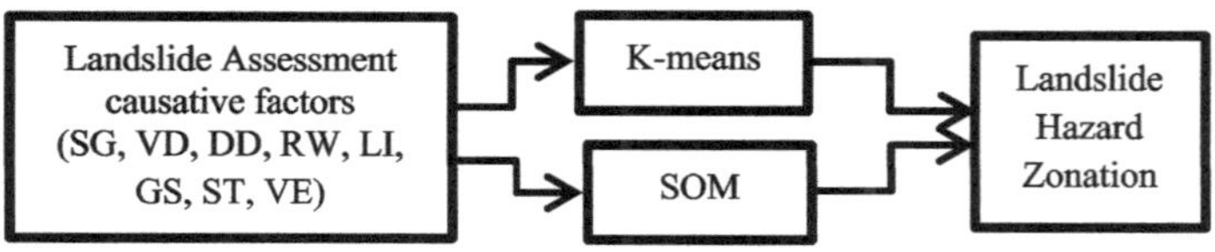

Fig. 4. System architecture.

3. Simulation and Results

This part presents the output of two clustering algorithms. Table 2 shows the number of identified landslide areas according to corresponding classes. There are 90 identified areas in a steep slope, while 63 areas are in moderate vertical displacement. In drainage density, 69 areas are low in drainages while 82 identified areas are highly weathered.

In lithology, all mining areas are highly weathered and highly fractured Most of these are sandstone; terrace gravel; highly fractured and highly weathered siltstone; andesite porphyry; ultramafic; agglomerate; diorite; hydrothermal breccia with tension cracks/or landslide scarp. Likewise, in ground stability, active landslides are evident in all areas with presence of tension cracks, bulges, terracettes, and seepage. As such, is also an indirect indicator of slope stability, but this will also contribute to slope instability with a particular condition to landslide due to existing mining areas.

In soil types, there are 42 identified areas under the class of silts and clays (50% or more of material is smaller than No. 200 sieve size) with a liquid limit of 50% or greater. These are elastic silt and elastic silt with sand likewise in vegetation 80 identified areas are in tertiary growth.

Table 2. Identified small scale mining areas and its classes.

Indicators	Classes	Number of identified landslide areas	Indicators	Classes	Number of identified landslide areas
Slope Gradient	FU	0	Lithology	WF	0
	MS	3		MF	0
	HS	6		HF	0
	SP	90		VF	99
Vertical Displacement	SH	1	Ground Stability	ST	0
	MV	63		SC	0
	DP	26		IL	0
	VD	9		AL	99
Drainage Density	LD	69	Soil Type	SG	42
	MD	28		SL	31
	HD	0		SD	16
	DD	2		GV	10
Rate of Weathering	SW	0	Vegetation	PG	0
	MW	17		SG	19
	HW	82		TG	80
	CW	0			

3.1 *K-Means Simulation*

The zonation for landslide hazards is clustered into four (4) to represent low, moderate, high, and very high regions. The data was extracted from the (.csv) files. The data used for training is 40% taken from the landslide dataset and 60% percent for the testing since this algorithm include as many samples of data as possible to obtain more accurate results.

The output from K-Means clustering program is provided in Table 3, which presents the numerical output data to interpret similarity and dissimilarity within clusters. The tested eight variables revealed four major clusters. It is shown in cluster 1 that higher value of soil type will contribute to higher susceptibility rate of vegetation and rate of weathering. This further concludes that the increase rate of soil type enhances the effect of weathering and vegetation. However, considering the slope gradient and drainage density, cluster 1 indicates that the soil type and the said variables inversely vary. It follows that those with a higher value of soil type has lesser value of slope gradient which also have a tendency of low drainage density.

While in cluster 3 vertical displacement has the highest value whereas the rates of weathering, soil type, and vegetation have the lowest value. This implies that vertical displacement indirectly varies from the rate of weathering, soil type, and vegetation. In effect, the area with high vertical displacement is the one with a lesser rate of weathering, soil type, and vegetation.

Table 3. Cluster centroids.

Variable	Cluster 1	Cluster 2	Cluster 3	Cluster 4	Grand centroid
SG	22.4286	64.3095	62.0833	38.9677	50.1818
VD	5.6071	8.9524	34.1667	8.6129	11.4293
DD	0.9286	1.6667	1.7500	2.0645	1.6970
RW	5.7857	5.3571	5.0000	5.7097	5.4848
LI	9.0000	9.0000	9.0000	9.0000	9.0000
GS	9.0000	9.0000	9.0000	9.0000	9.0000
ST	4.6429	2.5238	2.3333	3.6774	3.1616
VE	9.0000	7.8571	7.0000	8.8710	8.2323

As seen in Table 3, the vertical displacement obtained the highest mean of 34.1667 which is in Cluster 3 and centroid of 11.4293. In the same way, among all the factors, vertical displacement is the most diverse from its centroid, and so, it is the primary factor that affects the landslide hazard susceptibility. Moreover, slope gradient obtained the highest mean of 64.3095 which is in Cluster 2 and centroid of 50.1818; this means that the slope gradient is the second most diverse from its centroid, and so, it is the secondary factor that affects the landslide hazard susceptibility.

3.2. *Self-Organizing Map Simulation*

Another approach that aids in the interpretation of the maps provided particular external information like class labels is to plot the labels on the organized map. The distribution of the samples of each class, plotted on the map may also help in the interpretation process.

Same with K-Means, SOM used data extracted from .csv files. The model was trained by using 30% of the landslide data and 70% for the testing of data. After running the data through the SOM, the landslide areas were placed on a 10 x 10 grid representing their relative similarities.

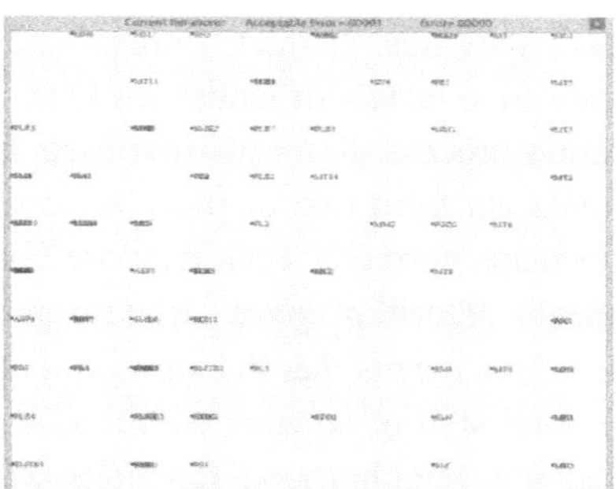

Fig. 5. Simulation of SOM algorithm.

Figure 5 is a representation of SOM visual output which determines the clustering of the data. There are 16 barangays from the different six towns of the province engage in SSM activities. Ninety-nine landslide areas observed during the comprehensive survey. Cansayong Slide 1 has been pushed to the top right of the highest mean in all of the eight causative factors, and this probably results to most elevated affected landslide area. While Layab Slide 6 has a decreased in mean value, has been pushed to the bottom left. Directly below on the bottom right-hand corner, Tugonan Slide 1 obtained the highest score in drainage density. The remaining neurons live between these extremes, with a landslide hazard zone occupying the center ground. Results compared with training and testing were accurate since it can map grouped items together and a number of zones have formed. Figure 6 point out the geographic location of the landslides identified areas in SSM.

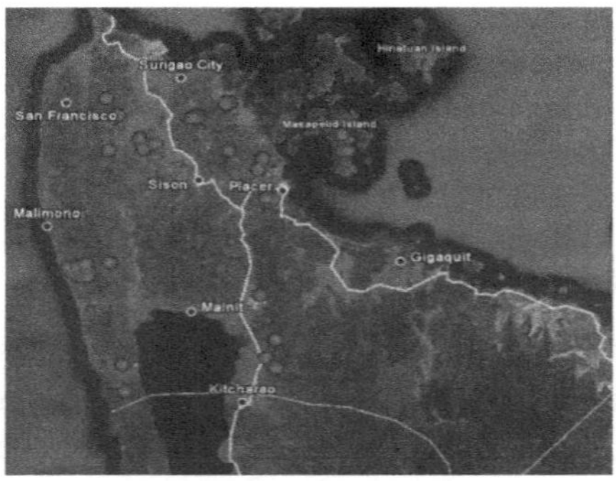

Fig. 6. Geographic location of the identified landslide areas in small scale mining.

Findings assert that the mining areas are in high to very high probability of landslide occurrence. These identified causative factors of the landslide are naturally occurring. One or two factors may be present in an area, but unregulated SSM activities make the mining sites susceptible to landslides as consequence of miners' activities. In fact, unregulated SSM activities cause presence of the eight causative factors of landslides [24].

During the simulation process, K-means was used as an interaction model, reading the landslide datasets from two or more factors, and adjust the cluster number. The algorithm maps correctly even data are combined in discrete and ordinal form. The k-means algorithm gives different answers when initiated at different starting values. This means that the algorithm does not always find the minimum value of the total sum of squares within a cluster. While in SOM, it uses a more local rule, the neighborhood functions wherein locations of the neurons are tuned to become ordered, and a meaningful coordinate system for the input features created on the lattice. Thus, SOM forms the required topographic map of the input patterns. Neurons were visualized as changing in positions in the weight space as learning takes place. However, it caused overlapping when real data is used in SOM, since it can have the same rate. Moreover, a test was also conducted in which the output from the K-means was used to feed into the SOM environment. Findings will not match since these two algorithms concentrate on a distinct kind of computation.

4. Conclusion

This paper concludes that the model can map the areas and can be used as a tool to visualize the identified causative factors of landslide. The performance of the two clustering algorithms differed since K-Means is sensitive to initialization. It depends on the initializing patterns according to the statistical information calculated from the data. SOM provides a more robust learning, but visual inspection needs to be done carefully because it requires prior checking of topology preservation property of trained map before starting visual inspection. The performance of clustering algorithms depends on the topology of data. The advanced multivariate techniques using clustering method provide useful spatial prediction of landslides with great extent of precision. Performance analysis of the K-Means depends a lot on a priori knowledge (K) and is very stable while Self-Organizing Map has stability and convergence which assured the principle of self-ordering but slow and many iterations needed for convergence likewise computationally exhaustive.

References

1. F. Guzzetti, Landslide Hazard and Risk Assessment. Concepts, Methods and Tools for the Detection and Mapping of Landslides, for Landslide Susceptibility Zonation and Hazard Assessment, and for Landslide Risk Evaluation. https://itunes.apple.com/au/book/landslide-hazard-risk-assessment/id944699313?mt=13. (Earth Sciences, 2014).
2. S.D. Pardeshil, S.E. Autade and S.S. Pardeshi, Landslide hazard assessment: recent trends and techniques, Pardeshi *et al.* SpringerPlus. 2:523 doi:10.11 86/2193-1801-2-523. http://www.springerplus.com/content/2/1/523 (2013).
3. D.T. Bui, T.A. Tuan, H. Klempe, B. Pradhan and I. Revhaug, Spatial Prediction Models for Shallow Landslide Hazards: A Comparative Assessment of the Efficacy of Support Vector Machines, Artificial Neural

Networks, Kernel Logistic Regression, and Logistic Model Tree, (2015), http://link.springer.com/article/10.1007/s10346-015-0557-6.

4. F. Sdao, D.S. Lioi, S. Pascale, S. Caniani and I.M. Mancini, Landslide susceptibility assessment by using a neuro-fuzzy model: a case study in the Rupestrian heritage rich area of Matera, *Natural Hazards and Earth System Sciences*, 13, 395–407, (2013).
5. P. Kayastha, Application of fuzzy logic approach for landslide susceptibility mapping in Garuwa sub-basin, East Nepal, *Frontier of Earth Science* 6(4): 420–432, (2012).
6. P. Kayastha, S.M. Bijukchhen, M.R. Dhital and F. De Smedt, GIS Based Landslide Susceptibility Mapping using a Fuzzy Logic Approach: A Case Study from Ghurmi-Dhad Khola Area, Eastern Nepal, *Journal Geological Society of India*, Vol. 82, (September 2013), pp. 249-261.
7. M.J. Friedel, Modeling hydrologic and geomorphic hazards across post-fire landscapes using a self-organizing map approach, *Environmental Modeling & Software*, Volume 26, Issue 12, (December 2011), pp. 1660-1674.
8. A. Vella and C. Vella, Neural Networks, University of London International Programmes, (2009), www.londoninternational.ac.uk
9. G.F. Lin, T.C. Wang and L.H. Chen, A Forecasting Approach Combining Self-Organizing Map with Support Vector Regression for Reservoir Inflow during Typhoon Periods, *Advances in Meteorology*, Volume 2016, (2016), Article ID 7575126, 12 pages. http://www.hindawi.com/journals/amete/2016/7575 126/abs/.
10. I. Stamova, H. Akca and G. Stamov, Qualitative Analysis of Dynamic Activity Patterns in Neural Networks, *Journal of Applied Mathematics*, Volume Article ID 208517, 2 pages, (2010), http://dx.doi.org/10.1155/2011/208517. http://www.hindawi.com/ journals/jam/2011/208517/
11. S.A. Sayad, Self-Organizing Map (SOM), Bahen Centre Information Technology, Toronto, Ontario, Canada, (2010).
12. T. Russo, M. Scardi and S. Cataudella, Applications of Self-Organizing Maps for Ecomorphological Investigations through Early Ontogeny of Fish, PLoS ONE 9(1):e86646.doi:10.1371/ journal.pone. 0086646, (2014).
13. M.J. Friedel, Modeling Hydrologic and Geomorphic Hazards Across Post-Fire Landscapes Using a Self-organizing Map Approach, doi:10.1016/j.envsoft.2011.07.001, (2011), http://www.sciencedirect.com/science/article/pii/S1364815211001629.
14. A. Skupin, J.R. Biberstine and K. Borner, Visualizing the Topical Structure of the Medical Sciences: A Self-Organizing Map Approach, DOI: 10.1371/journal.pone.0058779, (2013).

15. A.S.B. Samma and R.A. Salam, Adaptation of K-Means Algorithm for Image Segmentation, World Academy of Science, Engineering and Technology 50, (2009). http://citeseerx.ist.psu.edu/viewdoc/download?doi=10.1.1.193. 2635 &rep=rep1&type=pdf.
16. P.V. Gorservski, P.E. Gessler and P. Jankowski, Integrating a fuzzy k-means classification and a Bayesian approach for spatial prediction of landslide hazard, *The Journal of Geographical Systems*, (2003), 5:223-251DOI: 10.1007/s10109-003-0113-0, Retrieved from http://link.springer.com/article/
17. T.R. Martha, N. Kerle, C.J. van Westen, V. Jetten and K.V. Kumar, Segment Optimization and Data-driven Thresholding for Knowledge-based Landslide Detection by Object-based Image Analysis, Geoscience and Remote Sensing, IEEE Transactions on (Volume:49, Issue: 12), (2011), DOI: 10.1109/TGRS. 2011.2151866
18. IRINnews.org, PHILIPPINES: Artisanal mining compounds landslide risk, http://www.irinnews.org/report/94629/philippines-artisanal-mining-compounds-landslide-risk, (2012).
19. M.A. dela Cerna and E.A. Maravillas, Landslide Hazard GIS-based Mapping Using Mamdani Fuzzy Logic in Small Scale Mining Areas of Surigao del Norte, Philippines, Lecture Notes in Engineering and Computer Science: Proceedings of The World Congress on Engineering and Computer Science 2015, WCECS 2015, 21-23 October, (2015), San Francisco, USA, pp. 901-906.
20. T. Hentschel, F. Hruschka, M. Priester, Global Report on Artisanal and Small-Scale Mining, (2002), http//pubs.iied.org/pdfs/G00723.
21. D. Petley, Mining-related landslides in the Philippines, http://blogs.agu.org/ landslideblog/2011/04/28/mining-related-landslides-in-the-philippines/ 2011, (2011).
22. Unified Soil Classification System, http://gozips. uakron .edu/~mcbelch/documents/ UnifiedSoilClassification.pdf.
23. R.A. Mayer, T.S. Aziz and A.G. Rauber, Visualizing Class Distribution on Self-Organizing Maps, (2011).
24. M.A. dela Cerna and E.A. Maravillas, An Application of Partitive Clustering Algorithm for Landslide Hazard Zonation, Lecture Notes in Engineering and Computer Science: Proceedings of The International MultiConference of Engineers and Computer Scientists 2016, IMECS 2016, 16-18 March, 2016, Hong Kong, pp. 25-30.

EFFECTIVE LEARNING ALGORITHM FOR FUZZY INFERENCE SYSTEMS USING NEURAL GAS NETWORK

HIROFUMI MIYAJIMA*,‡, NORITAKA SHIGEI† and HIROMI MIYAJIMA†

*Graduate School of Biomedical Sciences, Nagasaki Univercity, Nagasaki, Japan

†Graduate School of Science and Engineering, Kagoshima University Kagoshima, Japan

‡k3768085@kadai.jp

It is known that learning methods of fuzzy inference systems using vector quantization (VQ) and steepest descend method (SDM) are superior in terms of the number of rules. Some learning methods to determine initial assignment of center parameters in antecedent part for membership functions. However, little learning methods determining the initial assignment of weight parameters in consequent part have been proposed. On the other hand, it has been shown that a learning method of radial basis function (RBF) networks using VQ and generalized inverse method (GIM) is effective. In this paper, we propose a new learning method using Neural Gas Network(NG) as VQ, GIM and SDM. The method iterates three stages in the outer loop of the algorithm. The first stage adjust the fuzzy rule arrangement by using NG, the second one determines the weights of fuzzy rules by using GIM, and the third one updates both of the rule arrangement and the weights. In order to demonstrate the effectiveness of the proposed method, numerical simulations for function approximation and pattern classification problems are performed. Specifically, it is shown that the proposed method reduces the learning time to about one-tenth compared to conventional methods in function approximation problem.

Keywords: Fuzzy Inference Systems; Vector Quantization; Neural Gas Network; Generalized Inverse Method; Appearance frequency.

1. Introduction

Many studies on self-tuning fuzzy systems have been made.[1–4] Their aim is to construct automatically fuzzy systems from learning data. Although most of the methods are based on steepest descend method (SDM), the obvious drawbacks of the method are its large time complexity and getting stuck in a shallow local minimum. Further, there is difficulty for learning with high dimensional spaces.[3–5] In order to overcome them, some novel methods have been developed, which 1) create fuzzy rules one by one starting from any number of rules, or delete fuzzy rules one by one starting from a sufficiently large number of rules,[5] 2) use GA (Genetic Algorithm) and PSO (Particle Swarm Optimization) to determine the structure of the fuzzy model,[6] 3) use fuzzy inference systems composed of small number of input rule modules, such as SIRMs (Single Input Rule Modules) and DIRMs (Double Input Rule Modules) methods,[8,9] and 4) use a self-organization or a vector quantization technique to determine the initial assignment.[7,10,11] Specifically, learning methods using VQ and SDM are superior in the number of rules compared with other learning methods. Most of conventional learning methods using VQ are ones that determine initial assignment of parameters for membership functions in antecedent part using only input part of learning data. Therefore, we proposed some learning methods to determine initial assignment of center parameters for membership functions in antecedent part using all learning data.[15] However, little learning methods determining the initial assignment of weight parameters in consequent part have been proposed.[16,17] On the other hand, it is known that a learning method of radial basis function (RBF) networks using VQ and generalized inverse method (GIM) is much effective.[2] In RBF networks, center and width parameters are determined using VQ in advance and weight parameters are determined using them. In this paper, we propose an effective learning method using VQ(NG), GIM and SDM using not only the parameters of membership function in antecedent part but also weight parameters in consequent part. The method consists of three stages iterated in the outer loop of the algorithm. Each of three stages utilizes either VQ, GIM or SDM for tuning the fuzzy inference system. The first stage adjusts the fuzzy rule arrangement by using VQ, the second one adjusts the weights of fuzzy rules by using GIM, and the last one adjusts both of the rule arrangement and the weights. In order to demonstrate the effectiveness of the proposed method, numerical simulations for function approximation and pattern classification problems are performed.

2. Preliminaries

2.1. *The conventional fuzzy inference model*

The conventional fuzzy inference model using SDM is described.[1,2] Let $Z_j = \{1, \ldots, j\}$ and $Z_j^* = \{0, 1, \ldots, j\}$ for the positive integer j. Let R be the set of real numbers. Let $\boldsymbol{x} = (x_1, \ldots, x_m)$ and y^r be input and output data, respectively, where $x_i \in R$ for $i \in Z_m$ and $y^r \in R$. Then the rule of simplified fuzzy inference model is expressed as

$$R_j \ : \ \text{if } x_1 \text{ is } M_{1j} \text{ and } \cdots \text{ and } x_m \text{ is } M_{mj} \text{ then } y \text{ is } w_j, \qquad (1)$$

where $j \in Z_n$ is a rule number, $i \in Z_m$ is a variable number, M_{ij} is a membership function of the antecedent part, and w_j is the weight of the consequent part.

A membership value of the antecedent part μ_j for input $\boldsymbol{x}$ is expressed as

$$\mu_i = \prod_{j=1}^{m} M_{ij}(x_j). \tag{2}$$

Let c_{ij} and b_{ij} denote the center and the width values of M_{ij}, respectively. If Gaussian membership function is used, then M_{ij} is expressed as follow:

$$M_{ij}(x_j) = \exp\left(-\frac{1}{2}\left(\frac{x_j - c_{ij}}{b_{ij}}\right)^2\right). \tag{3}$$

The output y^* of fuzzy inference is calculated by Eq. (4).

$$y^* = \frac{\sum_{i=1}^{n} \mu_i \cdot w_i}{\sum_{i=1}^{n} \mu_i}. \tag{4}$$

In order to construct the effective model, the conventional learning is introduced. The objective function E is determined to evaluate the inference error between the desirable output y^r and the inference output y^*.

In this section, we describe the conventional learning algorithm.[2]

Let $D = \{(x_1^p, \ldots, x_m^p, y_p^r) | p \in Z_P\}$ and $D^* = \{(x_1^p, \ldots, x_m^p) | p \in Z_P\}$be the set of learning data and the set of input data of D, respectively. The objective of learning is to minimize the following mean square error(MSE):

$$E = \frac{1}{P} \sum_{p=1}^{P} (y_p^* - y_p^r)^2. \tag{5}$$

In order to minimize the objective function E, each parameter $\alpha \in \{c_{ij}, b_{ij}, w_j\}$ is updated based on SDM as follows:[1,2]

$$\alpha(t+1) = \alpha(t) - K_\alpha \frac{\partial E}{\partial \alpha} \tag{6}$$

where t is iteration time and K_α is a constant. When the Gaussian membership function is used as the membership function, the following relation holds.

$$\frac{\partial E}{\partial c_{ij}} = \frac{\mu_j}{\sum_{j=1}^{n} \mu_j} \cdot (y^* - y^r) \cdot (w_j - y^*) \cdot \frac{x_j - c_{ij}}{b_{ij}^2} \tag{7}$$

$$\frac{\partial E}{\partial b_{ij}} = \frac{\mu_j}{\sum_{j=1}^{n} \mu_j} \cdot (y^* - y^r) \cdot (w_j - y^*) \cdot \frac{(x_j - c_{ij})^2}{b_{ij}^3} \tag{8}$$

$$\frac{\partial E}{\partial w_j} = \frac{\mu_j}{\sum_{j=1}^{n} \mu_j} \cdot (y^* - y^r) \tag{9}$$

The conventional learning algorithm is shown as Fig. 1,[1,2] where θ and T_{max} are therthold and the maximum number of learning, respectively. Note that the method is generative one. The method is called learning algorithm A.

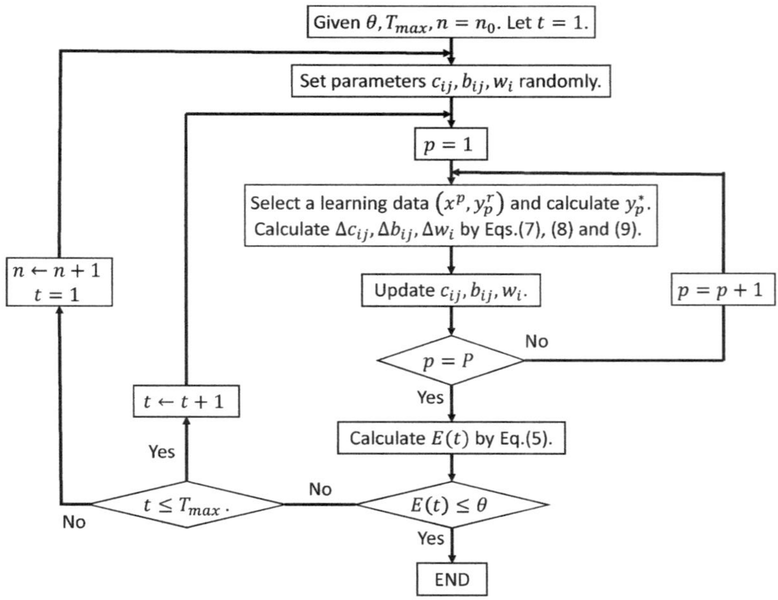

Fig. 1. The flowchart of the conventional learning algorithm.

2.2. *Neural Gas and K-means methods*

Vector quantization techniques encode a data space, e.g., a subspace $V \subseteq R^m$, utilizing only a finite set $C = \{\boldsymbol{c}_i | i \in Z_r\}$ of reference vectors (also called cluster centers), where m and r are positive integers.

Let the winner vector $\boldsymbol{c}_{i(\boldsymbol{v})}$ be defined for any vector $\boldsymbol{v} \in V$ as follows:

$$i(\boldsymbol{v}) = \arg\min_{i \in Z_r} ||\boldsymbol{v} - \boldsymbol{c}_i|| \tag{10}$$

From the finite set S, V is partioned as follows:

$$V_i = \{\boldsymbol{v} \in V | ||\boldsymbol{v} - \boldsymbol{c}_i|| \leq ||\boldsymbol{v} - \boldsymbol{c}_j|| \ for \ j \in Z_r\} \tag{11}$$

The evaluation function for the partition is defined as follows:

$$E = \sum_{i=1}^{r} \sum_{\boldsymbol{v} \in V_i} ||\boldsymbol{v} - \boldsymbol{c}_{i(\boldsymbol{v})}||^2 \tag{12}$$

For neural gas method(network),[13] the following method is used:

Given an input data vector $\boldsymbol{v}$, we determine the neighborhood-ranking $\boldsymbol{c}_{i_k}$ for $k \in Z^*_{r-1}$, being the reference vector for which there are k vectors $\boldsymbol{c}_j$ with

$$||\boldsymbol{v} - \boldsymbol{c}_j|| < ||\boldsymbol{v} - \boldsymbol{c}_{i_k}|| \tag{13}$$

If we denote the number k associated with each vector $\boldsymbol{c}_i$ by $k_i(\boldsymbol{v},\boldsymbol{c}_i)$, then the adaption step for adjusting the $\boldsymbol{c}_i$'s is given by

$$\triangle \boldsymbol{c}_i = \varepsilon \cdot h_\lambda(k_i(\boldsymbol{v}, \boldsymbol{c})) \cdot (\boldsymbol{v} - \boldsymbol{c}_i) \tag{14}$$

$$h_\lambda(k_i(\boldsymbol{v}, \boldsymbol{c})) = \exp(-k_i(\boldsymbol{v}, \boldsymbol{c})/\lambda) \tag{15}$$

where $\varepsilon \in [0, 1]$ and $\lambda > 0$. The number λ is called decay constant.

If $\lambda \to 0$, Eq. (14) becomes equivalent to the K-means method.[13] Otherwise, not only the winner $\boldsymbol{c}_{i_0}$ but the second, third nearest reference vector $\boldsymbol{c}_{i_1}$, $\boldsymbol{c}_{i_2}$, etc., are also updated.

Let $p(\boldsymbol{v})$ be the probability distribution of data vectors for V. The algorithm of the conventional neural gas method is shown in the following, where ε_{int}, ε_{fin}, θ and T_{max} are learning constants, threshold and the maximum number of learning, respectively. The method is called learning algorithm NG.

Learning Algorithm NG[13,15]

Step 1: The initial values of reference vectors are set randomly. The learning coefficients ε_{int} and ε_{fin} are set. Let T_{max} be the maximum number of learning time. Let $t = 1$.

Step 2: Given a data $\boldsymbol{v} \in V$ with $p(\boldsymbol{v})$, the neighborhood-ranking $k_i(\boldsymbol{v}, \boldsymbol{w}_i)$ is determined based on Eq. (13).

Step 3: Each reference vector $\boldsymbol{w}_i$ for $i \in Z_r$ is updated based on Eq. (14), where the learning coefficient $\varepsilon(t)$ is given as follows:

$$\varepsilon(t) = \varepsilon_{init} \left(\frac{\varepsilon_{fin}}{\varepsilon_{init}} \right)^{\frac{t}{T_{max}}} \tag{16}$$

Step 4: If $t = T_{max}$, then the algorithm terminates and the set $W = \{\boldsymbol{w}_i | i \in Z_r\}$ of reference vectors for V is obtained. Otherwise go to Step 2 as $t \leftarrow t + 1$.

If the data distribution $p(\boldsymbol{v})$ is not given in advance, a stochastic sequence of input data $\boldsymbol{v}(1), \boldsymbol{v}(2), \ldots$ which is based on $p(\boldsymbol{v})$ is given.[13]

By using Learning Algorithm NG, learning method of fuzzy systems is shown as follows:[12] In this case, assume that the distribution of learning data $\boldsymbol{D}^*$ is discrete uniform one. Let n_0 be the initial number of rules.

Learning Algorithm B

Step B1: For learning data D^*, Learning Algorithm NG is performed by using D^* as the set V. As a result, the set $\boldsymbol{c}$ of inference vectors for D^* is made, where $|\boldsymbol{c}| = n_0$.

Step B2: Each initial value c_{ij} is set to a reference vector.

Let

$$b_{ij} = \frac{1}{m_i} \sum_{\boldsymbol{x}_k \in C_i} (c_{ij} - x_{kj})^2, \tag{17}$$

where C_i and m_i are set of element and the number of learning data belonging to the i-th cluster C_i. Each initial weight w_i is selected randomly.

Step B3: The Steps of learning algorithm A without setting initial parameters of c_{ij}, b_{ij} and w_i are performed.

2.3. *Determination of weights using the generalized inverse method*

Let us explain fuzzy inference systems and interpolation problem using the generalized inverse method.[2] This problem can be stated mathematically as follows:

Given P points $\{\boldsymbol{x}^p = (x_1^p, \ldots, x_m^p)\,|p \in Z_P\}$ and P real numbers $\{y_p^r | p \in Z_P\}$, find a function $f : R^m \rightarrow R$ such that the following conditions are satisfied:

$$f(\boldsymbol{x}^p) = y_p^r \ (p \in Z_P) \tag{18}$$

In the case of fuzzy inference system, this problem is solved as follows :

$$y_p = f(\boldsymbol{x}^p) = \sum_{i=1}^{n} w_i \phi_{pi}(||\boldsymbol{x}^p - \boldsymbol{c}_i||) \tag{19}$$

$$\phi_{pi}(||\boldsymbol{x}^p - \boldsymbol{c}_i||) = \frac{\mu_i(\boldsymbol{x}^p)}{\sum_{l=1}^{n} \mu_l(\boldsymbol{x}^p)} \tag{20}$$

that is,

$$\Phi \boldsymbol{w} = \boldsymbol{y} \tag{21}$$

where $\Phi = (\phi_{pi})$ for $p \in Z_P$ and $i \in Z_n$ is a matrix.

Let $P = n$ and $\boldsymbol{x}^i = \boldsymbol{c}_i$. The width parameters are determined by Eq. (17). Then, if $\phi_i(\cdot)$ is suitably selected as Gaussian function, then the solution of weights $\boldsymbol{w}$ is obtained as

$$\boldsymbol{w} = \Phi^{-1}\boldsymbol{y} \tag{22}$$

Let us consider the case $n < P$. This is the realistic case. The optimal solution $\boldsymbol{w}^*$ that minimizes $E = ||y^r - \Phi\boldsymbol{w}||^2$ can be obtained as follows:

$$\boldsymbol{w}^+ = \Phi^T\boldsymbol{y} \text{ and } E_{min} = ||(I - \Psi)\boldsymbol{y}||^2 \tag{23}$$

where $\Phi^+ \triangleq [\Phi^T\Phi]^{-1}\Phi^T$, $\Psi \triangleq \Phi\Phi^T$ and I is identify matrix of $P \times P$.

Φ^+ is called the generalized inverse of Φ. The method using Φ^+ to determine the weights is called the generalized inverse method(GIM).[2]

[Example 1]

Let us consider a two-variable XOR problem using a fuzzy inference system with following rules:

$$R_i : \textit{if } x_1 \textit{ is } M_{i1} \textit{ and } x_2 \textit{ is } M_{i2} \textit{ then } y \textit{ is } w_i \tag{24}$$

where $i \in Z_4$.

Then, the fuzzy inference system with four fuzzy rules is assumed to be of the form:

$$y = \sum_{i=1}^{4} \frac{\mu_i w_i}{\sum_{l=1}^{4} \mu_l} \tag{25}$$

The four binary input vectors and the corresponding output vector are $\boldsymbol{x}_1 = (0,0)^T$, $\boldsymbol{x}_2 = (0,1)^T$, $\boldsymbol{x}_3 = (1,0)^T$, $\boldsymbol{x}_4 = (1,1)^T$ and $\boldsymbol{y} = (0,1,1,0)^T$, respectively. Let the center and width parameters of membership functions be $\boldsymbol{c}_1 = \boldsymbol{x}_1$, $\boldsymbol{c}_2 = \boldsymbol{x}_2$, $\boldsymbol{c}_3 = \boldsymbol{x}_3$, $\boldsymbol{c}_4 = \boldsymbol{x}_4$ and $b_{ij} = 0.25$ for $i \in Z_4$ and $j \in Z_2$, respectively.

In this case, the interpolation matrix Φ may be calculated as

$$\Phi = \begin{bmatrix} 0.258 & 0.250 & 0.250 & 0.242 \\ 0.250 & 0.258 & 0.242 & 0.250 \\ 0.250 & 0.242 & 0.258 & 0.250 \\ 0.242 & 0.250 & 0.250 & 0.258 \end{bmatrix} \tag{26}$$

As a result, the weight vector $\boldsymbol{w} = (w_1, w_2, w_3, w_4)^T = (20000, -20060.5, -20060.5, 20125)^T$ is obtained.

2.4. *The appearance frequency of input data based on the rate of change of output*

Learning Algorithm B is a method that determines the initial assignment of fuzzy rules by vector quantization using the set D^* of input for learning data. In this case, the set of output in learning data D is not used to determine the initial assignment of fuzzy rules. In the previous paper, Kishida proposed a method considering both input and output data to determine the initial assignment of fuzzy rules.[11]

Based on the Ref.11, the appearance frequency(probability) is defined as follows : Let D and D^* be the sets of learning data defined in Sec. 2.1.

Calculation Algorithm for the appearance frequency

Step 1: Give an input data $\boldsymbol{x}_i \in D^*$, we determine the neighborhood-ranking $(\boldsymbol{x}^{i_0}, \boldsymbol{x}^{i_1}, \ldots, \boldsymbol{x}^{i_k}, \ldots, \boldsymbol{x}^{i_{P-1}})$ of the vector $\boldsymbol{x}^i$ with $\boldsymbol{x}^{i_0} = \boldsymbol{x}^i$, $\boldsymbol{x}^{i_1}$ being closest to $\boldsymbol{x}^i$ and $\boldsymbol{x}^{i_k} (k = 0, \ldots, P-1)$ being the vector $\boldsymbol{x}^i$ for which there are k vectors $\boldsymbol{x}^j$ with $||\boldsymbol{x}^i - \boldsymbol{x}^j|| < ||\boldsymbol{x}^i - \boldsymbol{x}^{i_k}||$.

Step 2: Determine $H(\boldsymbol{x}^i)$ which shows the degree of change of inclination of the output around output data to input data $\boldsymbol{x}^i$, by the following equation:

$$H(\boldsymbol{x}^i) = \sum_{l=1}^{M} \frac{|y^i - y^{i_l}|}{||\boldsymbol{x}^i - \boldsymbol{x}^{i_l}||} \tag{27}$$

where $\boldsymbol{x}^{i_l}$ for $l \in Z_M$ means the l-th neighborhood-ranking of $\boldsymbol{x}^i$, $i \in Z_P$ and y^i and y^{i_l} are output for input $\boldsymbol{x}^i$ and $\boldsymbol{x}^{i_l}$, respectively. The number M means the range considering $H(\boldsymbol{x})$.

Step 3: Determine the appearance frequency $p_M(\boldsymbol{x}^i)$ for $\boldsymbol{x}^i$ by normalizing $H(\boldsymbol{x}^i)$.

$$p_M(\boldsymbol{x}^i) = \frac{H(\boldsymbol{x}^i)}{\sum_{j=1}^{P} H(\boldsymbol{x}^j)} \tag{28}$$

Learning algorithms C and D using the appearance frequency were proposed in Refs. 11, 15 and 16.

It is known that many rules are needed to palaces where output data change quickly for learning data. The appearance frequency $p_M(\boldsymbol{x})$ is a tool to find the optimum places for the assignment of fuzzy rules. The algorithm C is a heuristic method to find the optimum number of M and the algorithm D is a method to find it automatically.

3. The proposed method using GIM and NG

It is known that learning methods C and D using VQ and SDM is effective in accuracy and the number of rules to other methods.[13,15] Most of conventional learning methods using VQ are ones that determine initial assignment of center parameters in antecedent part for membership functions. However, little learning methods determining the initial assignment of weight parameters in consequent part have been proposed. On the other hand, it is known that learning method of RBF networks using VQ and GIM is much effective compared to other learning methods.[2] In RBF networks, center and width parameters are determined using k-means method as VQ in advance and weight parameters are determined. Therefore, we propose an effective learning method using NG, GIM and SDM using not only the parameters of membership function in antecedent part but also weight parameters in consequent part. It starts by breaking the method into three stages: learning intermediate stage adjusting the center and width parameters followed by updating the weight parameters using GIM. As for third stage, three parameters are updated by learning based on SDM. With iterating processes, parameters of the result of SDM are set to ones of the next process if the inference error is improved.

The proposed method is shown as follows:

Learning Algorithm E

Step 1: θ, T^0_{max}, T_{max}, M_0, M_{max} and β for $1 \leq M_{max}$ and $\beta < P$ are set. $M \leftarrow M_0$. Initial values of c^{best}_{ij}, b^{best}_{ij} and w^{best}_i are set randomly. Let E_{best} be the MSE for fuzzy inference system with c^{best}_{ij}, b^{best}_{ij} and w^{best}_i. $n \leftarrow n_0$.

Step 2: Let $c_{ij} \leftarrow c^{best}_{ij}$, $b_{ij} \leftarrow b^{best}_{ij}$, $w_i \leftarrow w^{best}_i$ and $t = 1$.

Step 3: Select a data $(\boldsymbol{x}^p, y^p)$ based on $p_M(\boldsymbol{x}^p, y^p)$ for $1 \leq p \leq P$.

Step 4: Update c_{ij} by Eq. (14).

Step 5: If $t < T^0_{max}$, go to Step 3 with $t \leftarrow t+1$, otherwise go to Step 6 with $t \leftarrow 1$.

Step 6: Determine b_{ij} by Eq. (17).

Step 7: Determine w_i by Eq. (23).

Step 8: Let $p = 1$.

Step 9: Given a data $(\boldsymbol{x}^p, y^r_p) \in \boldsymbol{D}$.

Step 10: Calculate μ_i and y^* by Eqs.(2) and (4).

Step 11: Update parameters c_{ij}, b_{ij} and w_{ij} by Eqs.(7), (8) and (9).

Step 12: If $p < P$ then go to Step 8 with $p \leftarrow p+1$.

Step 13: If $E > \theta$ and $t < T_{max}$ then go to Step 9 with $t \leftarrow t+1$, where E is computed as Eq. (5).

Step 14: If $E < E_{best}$ then $c^{best}_{ij} \leftarrow c_{ij}$, $b^{best}_{ij} \leftarrow b_{ij}$, $w^{best}_i \leftarrow w_i$ and $E_{best} \leftarrow E$.

Step 15: If $E > \theta$ and $M < M_{max}$ then go to Step 2 with $M \leftarrow M+\beta$, else if $E < \theta$ the algorithm otherwise terminate, otherwise go to Step 2 with $n \leftarrow n+1$, $M \leftarrow M_0$ and c^{best}_{ij}, b^{best}_{ij} and w^{best}_i are set randomly.

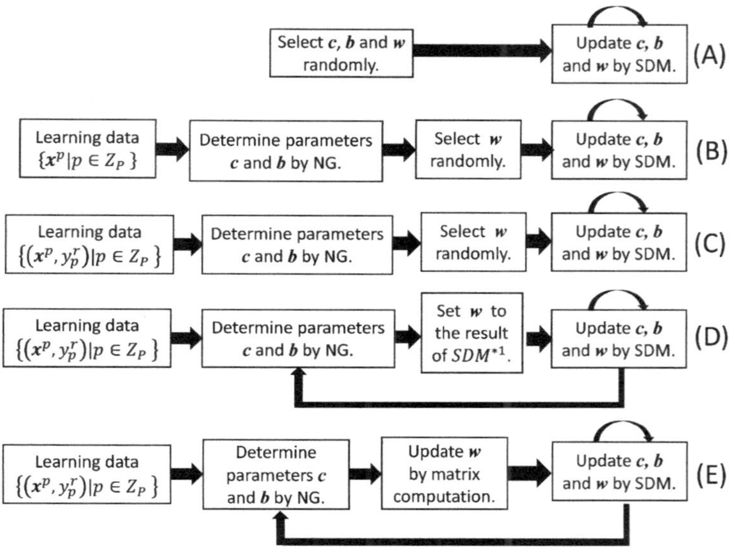

Fig. 2. Concept of conventional and proposed algorithms, where SDM and NG mean Steepest Descent Method and Neural Gas method. The mark $*1$ means that initial values of $\boldsymbol{w}$ are selected randomly.

In order to compare the proposed method with conventional ones, the following methods are introduced (See Fig. 2):

(A) Method A is one based on the algorithm of Fig. 1. Initial parameters of $\boldsymbol{c}$, $\boldsymbol{b}$ and $\boldsymbol{w}$ are set randomly and all parameters are updated using SDM until the inference error become sufficiently small.

(B) Method B is known as learning method of RBF networks.[2,10,12] Initial values of $\boldsymbol{c}$ are determined using D^* by VQ and $\boldsymbol{b}$ is computed using $\boldsymbol{c}$. Weight parameters $\boldsymbol{w}$ are randomly selected. Further, all parameters are updated using SDM until the inference error become sufficiently small.

(C) Method C was proposed in Ref. 10. Initial values of $\boldsymbol{c}$ are determined using D by VQ and $\boldsymbol{b}$ is computed using $\boldsymbol{c}$. Weight parameters are randomly selected. Further, all parameters are updated using SDM until the inference error become sufficiently small.

(D) Method D was proposed in Ref. 15. It is learning method composed of iterating two stages. The center parameters $\boldsymbol{c}$ are determined using D by VQ and $\boldsymbol{b}$ is computed using the center parameters. Weight parameters $\boldsymbol{w}$'s are set to the results of SDM, where the initial values of $\boldsymbol{w}$ are set randomly. Further, all parameters are updated using SDM for the definite number of learning time. With iterating processes, parameters of the result of SDM

are set to ones of the next process. Outer iterating process is repeated until the inference error become sufficiently small.
(E) Method E is the proposed one. It is learning method composed of iterating three stages. It starts by breaking the method into three stages: learning intermediate stage adjusting the center and width parameters followed by updating the weight parameters using the generalized inverse method. As for third stage, three parameters are updated using SDM for the definite number of learning time. With iterating processes, parameters of the result of SDM are set to ones of the next process.

4. Numerical simulations

In order to show the effectiveness of Learning Algorithm E, simulations of function approximation and classification problems are performed.

4.1. *Function approximation*

In order to show the effectiveness of Learning Algorithm E, numerical simulations of function approximation are performed. The systems are identified by fuzzy inference systems. This simulation uses four systems specified by the following functions with 4-dimensional input space $[0,1]^4$ (Eqs. (29) and (30)) and $[-1,1]^4$ (Eqs. (31) and (32)), and one output with the range $[0,1]$;

$$y = \frac{(2x_1 + 4x_2^2 + 0.1)^2}{37.21} \times \frac{(4\sin(\pi x_3) + 2\cos(\pi x_4) + 6)}{12} \tag{29}$$

$$y = \frac{(\sin(2\pi x_1) \times \cos(x_2) \times \sin(\pi x_3) \times x_4 + 1.0)}{2.0} \tag{30}$$

$$y = \frac{(2x_1 + 4x_2^2 + 0.1)^2}{74.42} + \frac{(3e^{3x_3} + 2e^{-4x_4})^{-0.5} - 0.077}{4.68} \tag{31}$$

$$y = \frac{(2x_1 + 4x_2^2 + 0.1)^2}{74.42} + \frac{(4\sin(\pi x_3) + 2\cos(\pi x_4) + 6)}{446.52} \tag{32}$$

The initial conditions for simulations are $K_c = 0.01$, $K_b = 0.01$, $K_c = 0.1$, $\varepsilon_{init} = 0.1$, $\varepsilon_{fin} = 0.01$, $\lambda = 0.7$, $T_{max} = 50000$ for Algorithms A, B and C and $T = 5000$ for Algorithms D and E. Further, the initial parameters are $\theta = 1.0\times10^{-4}$, $M = 300$, $M_0 = 300$, $\beta = 20$, $M_{max} = 480$, the number of learning and test data are 512 and 6400, respectively. In Figure 3, A, B, C, D and E mean Learning Algorithms A, B, C, D and E. Figure 3 shows the results for simulations. In Figure 3, the number of rules, MSE's for learning and test, and learning time(second) are shown, where the number

		Eq. (29)	Eq. (30)	Eq. (31)	Eq. (32)
A	the number of rules	4.2	13.6	7.2	5.1
	MSE for Learning($\times 10^{-4}$)	0.40	0.71	0.43	0.28
	MSE of Test($\times 10^{-4}$)	0.52	1.09	1.00	0.49
	learning time(s)	254.9	2617.5	777.9	365.7
B	the number of rules	5.6	14.9	5.2	3.7
	MSE of Learning($\times 10^{-4}$)	0.18	0.77	0.49	0.33
	MSE of Test($\times 10^{-4}$)	0.27	1.42	1.11	0.48
	learning time(s)	624.5	6766.1	513.0	255.5
C	the number of rules	4.8	15.6	5.5	4.0
	MSE of Learning($\times 10^{-4}$)	0.21	0.72	0.54	0.88
	MSE of Test($\times 10^{-4}$)	0.34	1.33	0.69	0.53
	learning time(s)	350.2	3125.0	447.6	210.0
D	the number of rules	3.0	8.4	4.0	3.0
	MSE of Learning($\times 10^{-4}$)	0.28	0.69	0.66	0.21
	MSE of Test($\times 10^{-4}$)	0.35	1.25	0.76	0.23
	learning time(s)	268.5	1673.3	425.9	267.0
E	the number of rules	3.0	8.0	8.6	5.5
	MSE of Learning($\times 10^{-4}$)	0.30	0.88	0.84	0.75
	MSE of Test($\times 10^{-4}$)	0.39	1.25	1.15	1.02
	learning time(s)	10.3	45.3	51.9	25.4

Fig. 3. The results for function approximation.

of rules means one when the threshold $\theta = 1.0\times 10^{-4}$ of inference error is achieved in learning. The result of simulation is the average value from twenty trials. As a result, the proposed method E reduces the learning time to about one-tenth compared to conventional methods.

4.2. *Two-category classification problems*

In the next, we perform two-category classification problems with spheres. In the classification problems, points on $[0,1] \times [0,1] \times [0,1]$ are classified into two classes: class 0 and class 1. The class boundaries are given as spheres centered at $(0.5, 0.5, 0.5)$. For Sphere, the inside of sphere with a radius of 0.3 is associated with class 1 and the outside with class 0. For Double-Sphere with radiuses of 0.2 and 0.4, the area between Spheres 1 and 2 is associated with class 1 and the other area with class 0. For triple-Sphere with radiuses of 0.1, 0.2 and 0.4, the inside of Sphere 1 and the area between Sphere 2 and Sphere 3 is associated with class 1 and the other area with class 0. The desired output y_p^r is set as follows: if $\boldsymbol{x_p}$ belongs to class 0, then $y_p^r = 0.0$. Otherwise $y_p^r = 1.0$. The numbers of learning and test data selected randomly are 512 and 6400, respectively. The simulation conditions are $K_c = 0.001$, $K_b = 0.001$, $K_w = 0.01$, $\varepsilon_{init} = 0.1$,$\varepsilon_{fin} = 0.01$, $\lambda = 0.07$ and $T_{max} = 50000$ for algorithms A, B and C and $T_{max} = 5000$ for algorithms D and E. The results on the rate of misclassification are shown in Fig.4. The result of simulation is the average from twenty trials. It is shown that the proposed algorithm E is superior in accuracy and the number of rules to other algorithms.

		Sphere	Double-Sphere	Triple-Sphere
A	the number of rules	6.0	17.3	14.7
	RM for Learning(%)	0.8	1.8	3.3
	RM of Test(%)	2.4	6.1	7.2
	learning time(s)	1423.7	6080.5	4201.5
B	the number of rules	5.3	24.6	10.7
	RM of Learning(%)	0.8	1.5	2.8
	RM of Test(%)	2.1	6.1	5.8
	learning time(s)	998.0	15894.7	4406.6
C	the number of rules	6.0	14.8	10.5
	RM of Learning(%)	0.8	1.8	2.9
	RM of Test(%)	2.2	5.7	6.1
	learning time(s)	1433.5	7835.5	4170.1
D	the number of rules	4.4	12.9	8.7
	RM of Learning(%)	0.8	1.8	2.7
	RM of Test(%)	2.1	5.5	5.7
	learning time(s)	1408.7	6417.7	3201.2
E	the number of rules	4.1	11.0	7.6
	RM of Learning(%)	0.7	1.6	3.1
	RM of Test(%)	1.9	5.3	5.7
	learning time(s)	1248.3	5134.3	2617.8

Fig. 4. The result for two-category classification problems.

5. Conclusion

In this paper, we proposed a new learning method composed of iterating three stages. It started by breaking the method into three stages: learning in the first stage, intermediate stage adjusting the center and width parameters, and the next stage of updating the weight parameters using the generalized inverse method (GIM). As the final stage, three parameters were updated by learning based on SDM. In order to demonstrate the effectiveness of the proposed method, numerical simulations for function approximation and pattern classification problems were performed. It was shown that the proposed method reduces the learning time compared to other methods in function approximation and is realized with high accuracy in small number of rules compared with other methods in classification problems. In the future work, we will propose more effective learning algorithm using VQ and the generalized inverse method compared to other methods.

References

1. C. Lin and C. Lee, Neural Fuzzy Systems, Prentice Hall, PTR (1996).
2. M.M. Gupta, L. Jin and N. Homma, Static and Dynamic Neural Networks, The Institute of Electrical and Electronics Engineers Press (2003).
3. J. Casillas, O. Cordon, F. Herrera and L. Magdalena, Accuracy Improvements in Linguistic Fuzzy Modeling, Studies in Fuzziness and Soft Computing, Vol. 129, Springer (2003).
4. B. Liu, Theory and Practice of Uncertain Programming, Studies in Fuzziness and Soft Computing, Vol. 239, Springer (2009).
5. S. Fukumoto, H. Miyajima, K. Kishida and Y. Nagasawa, A Destructive Learning Method of Fuzzy Inference Rules, Proceedings of The Institute of Electrical and Electronics Engineers on Fuzzy Systems, pp. 687-694 (1995).
6. O. Cordon, A historical review of evolutionary learning methods for Mamdani-type fuzzy rule-based systems, Designing interpretable genetic fuzzy systems, Journal of Approximate Reasoning, 52, pp. 894-913 (2011).
7. K. Kishida, H. Miya jima, M. Maeda and S. Murashima, A Self-tuning Method of Fuzzy Modeling using Vector Quantization, Proceding of FUZZ-IEEE'97, pp. 397-402 (1997).
8. N. Yubazaki, J. Yi and K. Hirota, SIRMS (Single Input Rule Modules) Connected Fuzzy Inference Model, Journal of Advanced Computational Intelligence, Vol. 1, Issue 1, pp. 23-30 (1997).
9. H. Miyajima, N. Shigei and H. Miyajima, Fuzzy Inference Systems Composed of Double-Input Rule Modules for Obstacle Avoidance Problems, IAENG International Journal of Computer Science, Vol. 41, Issue 4, pp. 222-230 (2014).

10. K. Kishida, H. Miyajima and M.Maeda, Destructive Fuzzy Modeling using Neural Gas Network, The Institute of Electronics, Information and Communication Engineers Transaction on Fundamentals, Vol. E80-A, No. 9, pp. 1578-1584 (1997).
11. K. Kishida and H. Miyajima, A Learning Method of Fuzzy Inference Rules using Vector Quantization, Procedings of the International Conference on Artificial Neural Networks, Vol. 2, pp. 827-832 (1998).
12. S. Fukumoto, H. Miyajima, N. Shigei and K. Uchikoba, A Decision Procedure of the Initial Values of Fuzzy Inference System Using Counterpropagation Networks, Journal of Signal Processing, Vol. 9, No. 4, pp. 335-342 (2005).
13. T. M. Martinetz, S. G. Berkovich and K. J. Schulten, Neural Gas Network for Vector Quantization and its Application to Time-series Prediction, The Institute of Electrical and Electronics Engineers Transaction on Neural Network, Vol. 4, No. 4, pp. 558-569 (1993).
14. W. Pedrycz and H. Izakian, Cluster-Centric Fuzzy Modeling, The Institute of Electrical and Electronics Engineers Transaction on Fuzzy Systems, Vol. 22, Issue 6, pp. 1585-1597 (2014).
15. H. Miyajima, N. Shigei, K. Kishida, Y. Akiyoshi and H. Miyajima, An Improved Learning Algorithm of Fuzzy Inference Systems using Vector Quantization, Advanced in Fuzzy Sets and Systems, Vol. 20, pp. 155-175 (2015).
16. H. Miyajima, N. Shigei and H. Miyajima, Fast Learning Algorithm for Fuzzy Inference Systems Using Vector Quantization, Lecture Notes in Engineering and Computer Science: Proceedings of The International MultiConference of Engineers and Computer Scientists 2016, International MultiConference of Engineers and Computer Scientists 2016, 16-18 March, 2016, Hong Kong, pp. 1–6.
17. H. Miyajima, N. Shigei and H. Miyajima, The Ability of Learning Algorithms for Fuzzy Inference Systems using Vector Quantization, International Conference on Neural Information Processing 2016, Lecture Notes in Computer Science 9950, pp. 479-488, Kyoto, October (2016).

ROAD USAGE RECOVERY IN MIYAGI PREFECTURE FOLLOWING THE 2011 TOHOKU EARTHQUAKE

NORIAKI ENDO

Iwate University, School of Humanities and Social Sciences
Morioka City, Iwate Prefecture, Tohoku District, Japan
n_endo@gw2k.hss.iwate-u.ac.jp

We evaluate regional differences in road recovery in Miyagi Prefecture following the 2011 Tohoku Earthquake. We divided Miyagi Prefecture into three areas, i.e., Inland, Northern Coastal, and Southern Coastal areas. According to the results of our study, we conclude that the recovery conditions of regional roads in different areas of Miyagi Prefecture following the 2011 Tohoku Earthquake differed. In the Northern Coastal area, 80% of the road distance was usable by April 15, 2011 and 90% was usable by May 27, 2011. In the Southern Coastal area, 80% of the road distance was usable by March 31, 2011 and 90% was usable by April 8, 2011. Recovery in the Southern Coastal area was much faster compared to that in the Northern Coastal area. We assume that this is due to the shape of the coastlines. The coastlines in the Northern coastal area are primarily rias. The coastlines in the Southern coastal area are mostly sandy. Furthermore, we have concluded that the recovery conditions of the regional roads following the 2011 Tohoku Earthquake in the Northern Coastal area of Miyagi Prefecture were similar to those in the Southern Coastal area of Iwate Prefecture. In the disaster regions, similar recovery conditions were found according to geographic positions and features.

Keywords: 2011 Tohoku Earthquake; big data analysis; G-GOOK telematics data; probe data; probe information analysis; vehicle tracking map

1. Introduction

1.1. *The 2011 Tohoku Earthquake*

The 2011 Tohoku Earthquake [Fig.1] struck the northeastern coast of Japan on March 11, 2011. Subsequently, the region was severely affected by the tsunami. Following these natural disasters, the electricity, water, and gas supplies were shut down in both coastal and inland areas.[1–4] Furthermore, the road travel was disrupted in many parts of the region.

1.2. *Purpose*

The primary purpose of our study was to evaluate the regional differences in road recovery in Miyagi Prefecture following the 2011 Tohoku Earthquake. Therefore, based on geographic position and features, we divided Miyagi Prefecture into Inland, Northern Coastal, and Southern Coastal areas. During the disaster, these areas were affected differently. For example, due to differences in coastal features, the tsunami struck the Northern Coastal area more heavily than the Southern Coastal area. Therefore, we assumed that there were specific differences among the three studied regions during the road recovery process following the disaster.

The secondary purpose of our study was to compare the regional differences with regard to road recovery in Miyagi Prefecture, the target of this study, compared to road recovery in Iwate Prefecture, which was evaluated in our previous study.[4]

2. Our previous studies

2.1. *Previous study 1*

Our previous study 1[1] was focused on the use of the main roads in the Southern Coastal area of Iwate Prefecture. The usable distances of the main roads following the 2011 Tohoku Earthquake have been calculated from the G-BOOK telematics data.[6]

The main findings of this study are as follows:

(1) The usable distance of the roads in a weekly period has continuously increased from March 18 to April 7, 2011, but it has fluctuated thereafter.

(2) Defining the cumulative usable distance up to September 30, 2011 as 100%, it has been determined that 80% of the road distance has been usable by April 7, 2011 and 90% by April 29, 2011.

(3) The use of the main road in the coastal area of Iwate Prefecture has been completely recovered by April 29, 2011.

2.2. *Previous study 2*

Our previous study 2[2,3] was focused on the use of the main roads not only in the Southern but also in the Northern Coastal area of Iwate Prefecture.

The cumulative usable road distance ratio of the main roads has been precisely calculated for each city using the free and open source geographical information system software QGIS. The main findings in this study[2,3] are listed below.

(1) The change in the cumulative usable road distance ratio during the research period differed from one city to the next.

(2) The ratio increases in the usable distances of Kuji, Iwaizumi, and Noda were extremely delayed.

In our study, we were able to determine related roads by analyzing the maps generated by the QGIS software. For Kuji and Iwaizumi, the road whose recovery was significantly delayed is the Iwate Prefectural Road number 7 (Kuji-Iwaizumi line). For Noda, the road whose recovery was significantly delayed is the Iwate Prefectural Road number 273 (Akka-Tamagawa line).

(3) In our previous study 1,[1] we determined that the use of the main road in the Southern Coastal area of Iwate Prefecture was completely recovered by April 29, 2011.

However, in this study, when we have precisely observed the change in the usable road distance ratio during the research period for each city, the ratio increase in the usable road distance of Kamaishi has been delayed compared with other Southern Coastal cities.

For Kamaishi City, the road whose recovery was significantly delayed is the Iwate Prefectural Road number 249 (Sakuratoge-Heita line).

2.3. *Previous study 3*

In our previous study,[4] we calculated the regional differences for road recovery in Iwate Prefecture following the 2011 Tohoku Earthquake. We divided Iwate Prefecture into four areas, i.e., Northern Inland, Southern Inland, Northern Coastal, and Southern Coastal areas. The main results of the previous study are as follows.

First, we determined that, for Northern and Southern Inland areas, 80% of the road distance was usable by April 15, 2011 and 90% by May 27, 2011, which indicates that the recovery speed in these areas was slightly slower than that in the Southern Coastal area.

Second, we found that, for the Northern Coastal area, 80% of the road distance was usable by April 29, 2011 and 90% by June 24, 2011, which implies that the recovery speed in the Northern Coastal area was significantly slower than that in the Southern Coastal area.

Hence, we concluded that these findings are related to the fact that road recovery efforts were more focused on regions heavily affected by the earthquake and tsunami.

2.4. *Current paper*

Current paper is the extended version of our conference paper.[5]

3. Telematics data and vehicle-tracking map

Telematics is a general term encompassing telecommunications and informatics. A telematics service provides various personalized information for users, especially for drivers of automobiles. G-BOOK is a telematics service provided by Toyota Motor Corporation.

To calculate the usable distance of the main roads, we applied the vehicle tracking map originally created by Hada *et al.*[7] after the 2007 Niigataken Chuetsu-oki earthquake.

That vehicle tracking map was based on telematics data provided by Honda Motor Company. Similarly, in our study, we used the vehicle tracking map based on telematics data provided by Toyota's G-BOOK system [Fig. 2].

Registered members of G-BOOK can access telematics services to acquire GPS data for car navigation systems and interactive driving data, such as traffic jam points, road closures, and weather reports.

Such comprehensive data acquisition is possible because the telematics system server receives accurate location data (geographic coordinates) from its registered members.

Telematics services are extremely useful to drivers. Because the accurate driving routes of registered users remain in the system server, they are accessible to traffic researchers in various fields.

4. Research Methods

4.1. *Research area*

The current study was focused on the entire area of Miyagi Prefecture (i.e., the Inland, Southern Coastal, and Northern Coastal areas) [Fig. 3].

4.2. *Research materials*

In our current study, we have used the vehicle tracking maps built from the G-BOOK telematics data that is available on the Internet on March 18, 2011 following the 2011 Tohoku Earthquake.[6]

The data used in this study have been collected between March 18 and September 30, 2011 (i.e., approximately six months following the 2011 Tohoku Earthquake).

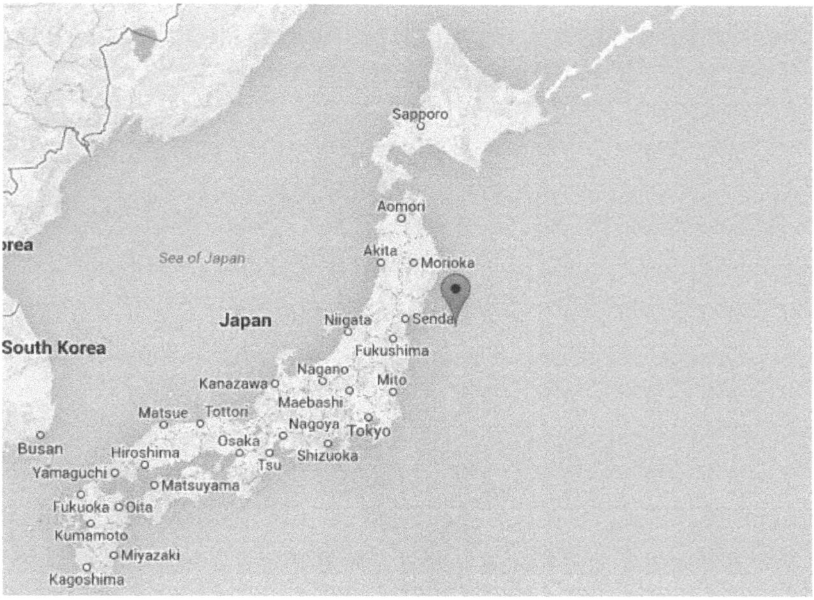

Fig. 1. Center of the 2011 Tohoku Earthquake, March 11, 2011 (https://www.google.co.jp/maps/).

4.3. *System*

Hardware:

The computations have been performed on a standard PC laptop with a Core i5-4200U CPU (1.6 GHz) and 4 GB memory (SONY VAIO PRO 11).

Software:

The software QGIS version 2.6.1[10] and LibreOffice Calc 4.2.7 spreadsheet software[11] running on the Windows 7 Professional operating system have been used in this study. It is well-known that QGIS is one of the most popular geographic information systems used worldwide.

Prior to the abovementioned applications for geographical data processing, we have used the ogr2ogr software[12] on the Linux operating system along with Vine Linux 4.2,[13] which is a Linux distribution developed by a Japanese Linux community.

Note that QGIS, LibreOffice Calc, ogr2ogr, and Vine Linux are open source software systems freely available on the Internet.

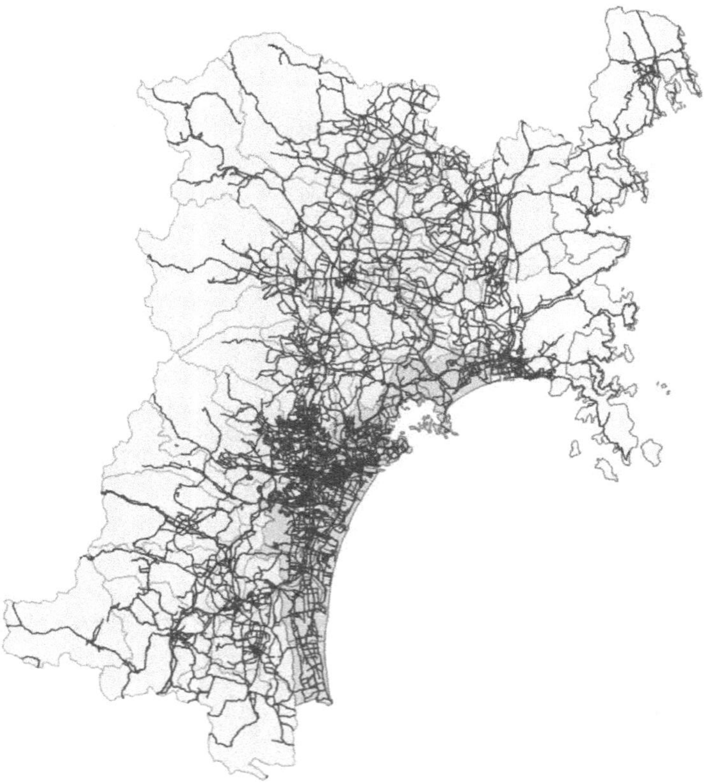

Fig. 2. Vehicle tracking map of Miyagi Prefecture.

4.4. *Data Processing*

(1) The vehicle tracking maps constructed from the G-BOOK telematics data have been provided in the Google map KMZ format. For our analysis, we have first converted the KMZ files to SHP files (i.e., shape-files), which are compatible with ArcGIS using the ogr2ogr software.

(2) Next, the data coordinates have been converted from the terrestrial latitude and longitude to the x and y coordinates in a rectangular coordinate system using the ogr2ogr software.

(3) To reduce the computation time, the data file has been clipped to small files containing only the research area using the QGIS software.

(4) After merging daily data into weekly data and removing duplicate data, we have been able to calculate the exact usable road distance available

Fig. 3. Miyagi Prefecture divided into Inland, Northern Coastal, and Southern Coastal areas. The perimeter of a city is shown by a gray polygon.

for a given week using the QGIS software.

In this context, a usable road is one on which at least one vehicle has been probed during the observation period.

The purpose of converting the daily data to weekly data was to smooth the daily fluctuations in the traffic flows.

(5) Next, we have calculated the proportion of the cumulative distance up to the specified date using the LIbreOffice Calc software. Note that the cumulative distance up to September 30, 2011 was considered 100%.

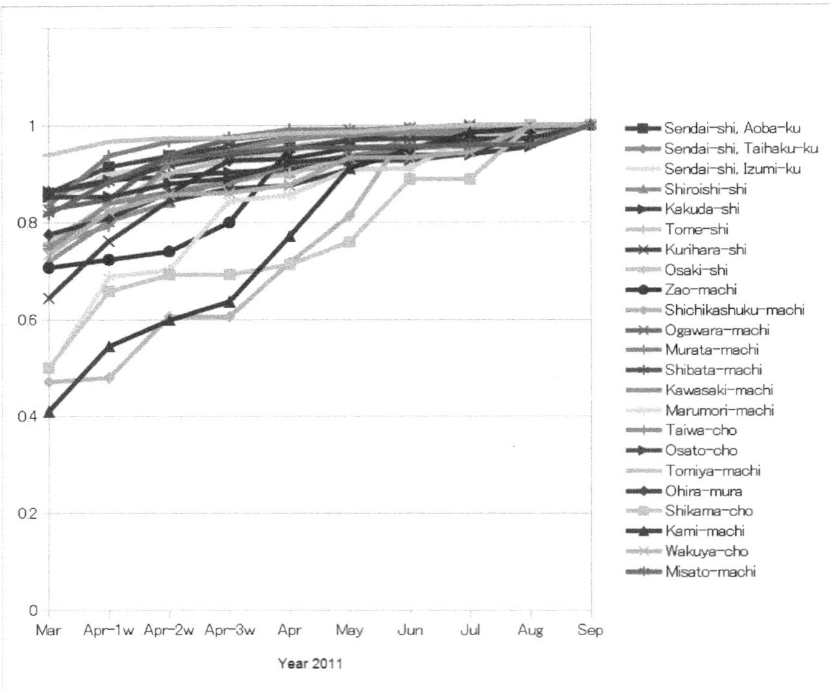

Fig. 4. Cumulative usable road distance ratio for the Inland area. The vertical scale displays the cumulative distance proportion of the usable roads (relative to the cumulative distance on September 30, 2011) for each date.

5. Results

5.1. *Regional road recovery differences*

Defining the cumulative usable distance up to September 30, 2011 as 100%, the percentages of usable road distances are given in Table 2. In Table 2, the upper lines indicate the cumulative usable road distances (in meters), and the lower lines represent the ratio of cumulative usable road distance.

(1) Inland area

It was determined that 80% of the road distance was usable by April 7, 2011 and 90% was usable by April 29, 2011. The recovery speed in Inland area was slightly slower than that in the Northern Coastal area.

(2) Northern Coastal area

It was determined that 80% of the road distance was usable by April 15, 2011 and 90% was usable by May 27, 2011.

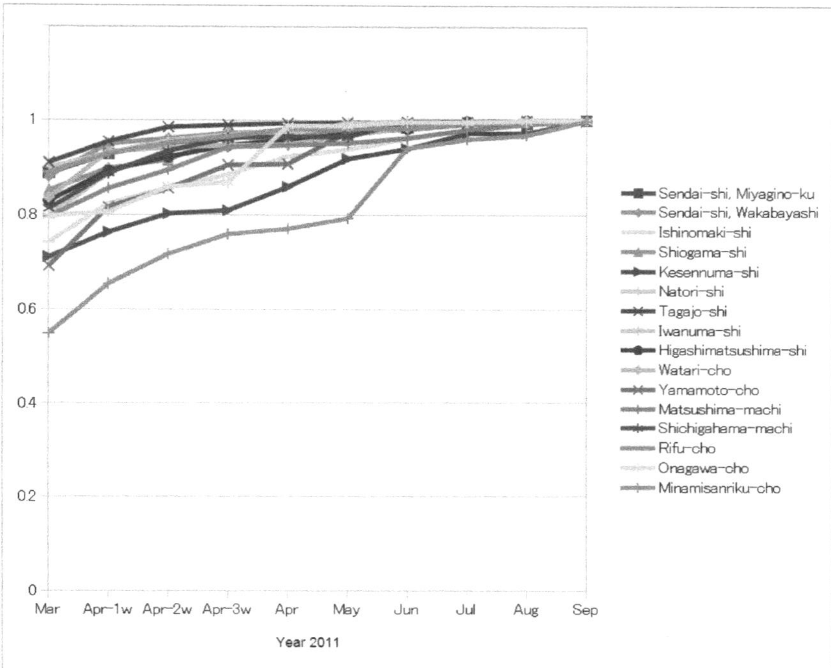

Fig. 5. Cumulative usable road distance ratio for the Coastal area. The vertical scale displays the cumulative distance proportion of the usable roads (relative to the cumulative distance on September 30, 2011) for each date.

(3) Southern Coastal area

It was determined that 80% of the road distance was usable by March 31, 2011 and 90% by April 8, 2011. Recovery rates in the Southern Coastal area were significantly higher compared to the Northern Coastal area.

5.2. *Cities where road recovery was delayed significantly*

(1) Inland area [Fig. 4]

The usable road distance ratio increases for Kami, Shichigashuku, Shikama, Marumori, and Zao towns were delayed significantly compared to those for other municipalities in the Miyagi Prefecture Inland area. We were able to identify related roads by analyzing maps generated using QGIS software.

Table 1. Regional differences for road recovery in Miyagi Prefecture (cumulative usable road distances (meters) and ratios).

	Mar 31	Apr 08	Apr 15	Apr 22	Apr 29
Whole	3710404.71	4070465.56	4302746.51	4432316.59	4555942.55
Inland	0.749	0.822	0.869	0.895	0.920
Northern	689499.47	760734.33	798356.91	820800.18	861329.64
Coastal	0.719	0.793	0.832	0.856	0.898
Southern	1401296.10	1501843.72	1544385.42	1576106.64	1589184.41
Coastal	0.853	0.914	0.940	0.959	0.967

	May 27	Jun 24	Jul 29	Aug 26	Sep 30
Whole	4689705.40	4784541.96	4853809.92	4898789.43	4951562.17
Inland	0.947	0.966	0.980	0.989	1.000
Northern	884760.76	917857.84	938069.52	946266.16	959149.12
Coastal	0.922	0.957	0.978	0.987	1.000
Southern	1610740.67	1621919.88	1634768.72	1639012.45	1643175.95
Coastal	0.980	0.987	0.995	0.997	1.000

For Kami Town, the roads whose recovery was significantly delayed are Miyagi Prefectural Road number 226 (Iwadeyama-Miyazaki line) and the road that runs between National Route 347 and Ishikawa Farm.

For Shichigashuku Town, the roads whose recovery was significantly delayed are Miyagi Prefectural Road number 13 (Kaminoyama-Shichigashuku line) and Miyagi Prefectural Road number 51 (the Minamizao-Shichigashuku line).

For Shikama Town, the roads whose recovery was significantly delayed are the road between Shin-Kitsupushi and Miyagi Prefectural Road number 156, two roads between Miyagi Prefectural Road number 156 and Takane, and the road that connects Miyagi Prefectural Road number 156, Nishihara, and Ohara Farm.

For Marumori Town, the roads whose recovery was significantly delayed are as follows.

-Miyagi Prefectural Road number 45 (Marumori-Ryozen line; near Hippo)

-Miyagi Prefectural Road number 228 (Soma-Ouchi line; near Aoba Hot Spring)

Table 2. Regional difference for road recovery in Iwate Prefecture (cumulative usable road distances (meters) and ratios).[4]

	Mar 31	Apr 08	Apr 15	Apr 22	Apr 29
Northern	1850873.51	2070475.04	2152283.40	2251706.55	2296780.33
inland	0.711	0.795	*0.827*	0.865	0.882
Southern	2386739.37	2722218.13	2870724.95	2991566.34	3074037.86
inland	0.689	0.785	*0.828*	0.863	0.887
Northern	394082.91	449373.30	471389.15	524694.94	564884.21
coastal	0.583	0.665	0.698	0.777	*0.836*
Southern	666655.53	779216.76	814157.35	821573.46	843427.24
coastal	0.713	*0.834*	0.871	0.879	*0.902*

	May 27	Jun 24	Jul 29	Aug 26	Sep 30
Northern	2403134.24	2483876.38	2532471.47	2571798.73	2603629.91
inland	*0.923*	0.954	0.973	0.988	1.000
Southern	3201569.09	3310978.41	3396994.03	3429255.89	3465835.57
inland	*0.924*	0.955	0.980	0.989	1.000
Northern	589275.55	614317.97	618734.27	668603.07	675390.65
coastal	0.872	*0.910*	0.916	0.990	1.000
Southern	855066.71	903310.83	910464.22	920120.72	934580.47
coastal	0.915	0.967	0.974	0.985	1.000

-The road between Miyagi Prefectural Road number 12 and Nenashifuji

-the road between Miyagi Prefectural Road number 12 and Magatake-Goanji

(2) Northern Coastal area [Fig. 5]

The ratio increases in the usable distance for Minami-Sanriku Town and Kesen-numa City were delayed significantly compared to those for other municipalities in the Miyagi Prefecture Coastal area. We were able to determine related roads by analyzing maps generated using QGIS software.

For Minami-Sanriku Town, the roads whose recovery was significantly delayed are as follows.

-Miyagi Prefectural Road number 225 (Tomarizaki-hanto line)

-The road to Kesenma-Oshima Jinjya (Shrine) from Miyagi Prefectural Road number 236 (Haraikawa-Machimukai line)

-The route that runs through National Route 398, Sakuraba-shukaijyo, Doujishita-shikinosato, and National Route 398

For Kesen-numa City, the roads whose recovery was significantly delayed are as follows.

-The road to Nakai and Karakuwa from Miyagi Prefectural Road number 26 (Kesennuma-Karakuwa line)

-The roads near Minami Kesennuma Station (Kesennuma Line of the Japan Railway Company) and Uchinowaki

-Miyagi Prefectural Road number 26 (two sections: near Ootougeyama and near Atagoyama)

-Miyagi Prefectural Road number 26 (the section from Nakai Elementary School to the southern end of the route near Misaki Shrine)

3) Southern Coastal area [Fig. 5]

The ratio increases in the usable distance of Matsushima Town was delayed significantly compared to those for other municipalities in the Miyagi Prefecture Coastal area. We were able to determine related roads by analyzing maps generated using QGIS software.

For Matsushima Town, the roads whose recovery was significantly delayed are as follows.

-The road from National Route 45 to Shintomi-yama

-The road from National Route 346 to the entrance of the tunnel near Hataya

-The road from Nakamachi to Shinden

6. Discussion

6.1. *Regional differences for the road recovery in Miyagi Prefecture*

In the current study, we have evaluated regional differences in road recovery in Miyagi Prefecture following the 2011 Tohoku Earthquake. We divided Miyagi Prefecture into three areas, i.e., Inland, Northern Coastal, and Southern Coastal areas. Based on our evaluation results, we have concluded that the recovery conditions of regional roads following the 2011 Tohoku Earthquake differed based on area. In the Northern Coastal area, it was determined that 80% of the road distance was usable by April 15, 2011 and 90% was usable by May 27, 2011. On the other hand, in the Southern Coastal area, it was determined that 80% of the road distance was usable by March 31, 2011 and 90% was usable by April 8, 2011. Recovery in the Southern Coastal area was significantly faster compared to the Northern Coastal area.

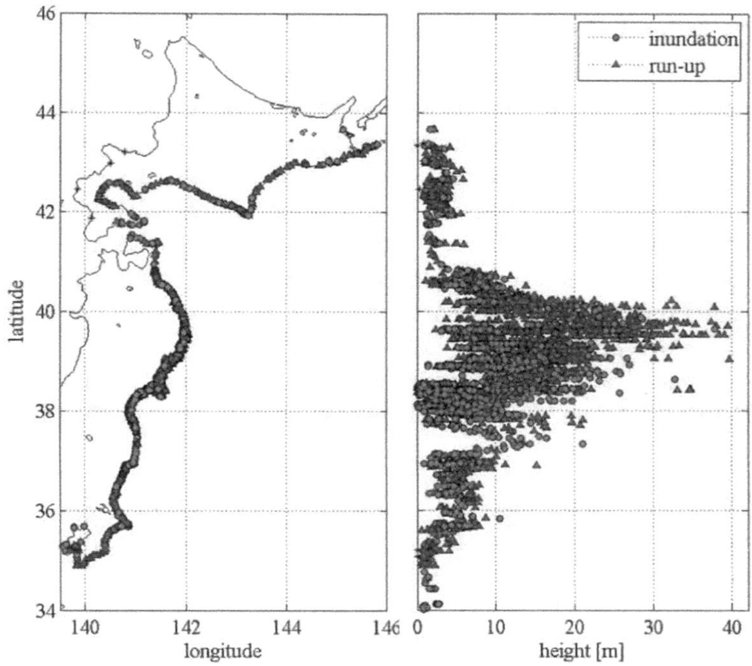

Fig. 6. Tsunami inundation following the 2011 Tohoku Earthquake. The left shows examination points. The right shows inundation and run-up (meters) of the points (cited from The 2011 Tohoku Earthquake Tsunami Joint Survey Group 2011[8]).

The initial cumulative usable road ratio (relative to cumulative distance on September 30, 2011) calculated on March 31 was 0.72 in the Northern Coastal area compared to 0.85 in the Southern Coastal area. The difference decreased until April 29, 2011. Nevertheless, it took considerable time before normal road conditions were restored in the Northern Coastal area.

We assume that this difference was related to differences in the height of the tsunami's inundation[8] [Fig. 6]. The coastlines observed in the Northern coastal area are primarily characterized by rias. On the other hand, the coastlines observed in the Southern coastal area are primarily sandy.

The coastal shape influences the height of the tsunami. Subsequently, the height of the tsunami influences the amount of damage to roads and land.

6.2. *Comparison of regional differences for road recovery in Miyagi Prefecture and Iwate Prefecture*

According to the results of our study, we have concluded that the recovery conditions of the regional roads following the 2011 Tohoku Earthquake in the Northern Coastal area of Miyagi Prefecture were similar to those in the Southern Coastal area of Iwate Prefecture [Tables 1 and 2].

We assume that this is because these regions border each other and their geographical features are similar. Furthermore, the recovery conditions of the regional roads following the 2011 Tohoku Earthquake in the Inland area of Miyagi Prefecture were similar to those in the Northern and Southern Inland area of Iwate Prefecture [Tables 1 and 2].

As discussed in this section, in the disaster regions, similar recovery conditions were found according to geographic positions and features.

Acknowledgments

We would like to thank all the people worldwide who assisted the people of Japan following the 2011 Tohoku Earthquake.

References

1. Hayato Komori and Noriaki Endo, "Road Distance Traveled by Vehicles Following the 2011 Tohoku Earthquake, Calculated by G-BOOK Telematics Data," *Proceedings of the 9th International Conference on Signal-Image Technology & Internet-Based Systems*, SITIS 9, Dec. 2-5, 2013, Kyoto, Japan, pp. 870-874.
2. Noriaki Endo and Hayato Komori, "Analysis of Vehicle Tracking Maps in Iwate Prefecture Following the 2011 Tohoku Earthquake, *Lecture Notes in Engineering and Computer Science - Proceedings of the International MultiConference of Engineers and Computer Scientists 2015*, IMECS 2015, Mar. 18-20, 2015, Hong Kong, pp. 124-128. Available at http://www.iaeng. org/publication/IMECS2015/IMECS2015_pp124-128.pdf
3. Noriaki Endo and Hayato Komori, "Municipal Difference on Road Recovery in Iwate Prefecture Following the 2011 Tohoku Earthquake," *IAENG Transactions on Engineering Sciences - Special Issue for the International Association of Engineers Conferences 2015*, October, 2016, World Scientific, pp. 226-235.
4. Noriaki Endo and Hayato Komori, "Regional Difference on Road Recovery in Iwate Prefecture Following the 2011 Tohoku Earthquake," *Lecture Notes in Engineering and Computer*

Science — Proceedings of the World Congress on Engineering and Computer Science 2015, WCECS 2015, Oct. 21-23, 2015, Berkeley, USA, pp. 162-167. Available at http://www.iaeng.org/publication/WCECS2015/WCECS2015_pp162-167.pdf

5. Noriaki Endo, "Regional Differences in Miyagi Prefecture Road Recovery Following the 2011 Tohoku Earthquake," *Lecture Notes in Engineering and Computer Science — Proceedings of The International MultiConference of Engineers and Computer Scientists 2016*, IMECS 2016, Mar. 16-18, 2016, Hong Kong, pp. 151-156. Available at http://www.iaeng.org/publication/ IMECS2016/IMECS2016_pp151-156.pdf
6. Vehicle Tracking Map Constructed by G-BOOK Telematics Data (In Japanese). http://g-book.com/pc/spot/Tohoku-Jishin.asp
7. Yasunori Hada, Takeyasu Suzuki, Hiroki Shimora, Kimiro Meguro and Noriko Kodama: Issues and Future Prospect on Practical Use of Probe Vehicle Data for Disaster Reduction -Provision of the Vehicle Tracking Map in the 2007 Niigataken Chuetsu-oki Earthquake- (In Japanese), *Journal of Japan Association for Earthquake Engineering*, 9(2), 148-159, 2009. Available at http:// www.jaee.gr.jp/stack/submit-j/v09n02/090211_paper.pdf
8. The 2011 Tohoku Earthquake Tsunami Joint Survey Group: Results of Tsunami Surveys (July 5, 2011). http://www.coastal.jp/tsunami2011/index. php?Field
9. Vehicle Tracking Map Constructed by G-BOOK Telematics Data. http://g-book.com/disasterMap/ALL/Tohoku-Jishin_YYYYMMDD.kmz. (For example, Tohoku-Jishin_20110318.kmz. The data files of tracking map are still available on January 30, 2016.)
10. Official website of QGIS, A Free and Open Source Geographic Information System. http://www.qgis.org/en/site/
11. Official Website of the LibreOffice Project. http://www.libreoffice.org/
12. Official website of GDAL: ogr2ogr. http://www.gdal.org/ogr2ogr.html
13. Official website of Vine Linux (In Japanese). http://www.vinelinux.org/

SCHEDULING ALGORITHM BASED ON GA FOR TASKS WITH TIME CONSTRAINTS

M. YOO* and T. YOKOYAMA†
Department of Computer Science, Tokyo City University,
1-28-1 Tamazutsumi, Setagaya-ku, Tokyo, Japan
**yoo@cs.tcu.ac.jp*
†yokoyama@cs.tcu.ac.jp

This paper proposes a new real-time task scheduling algorithm which considers the precedence relationship between tasks. In solution algorithms, the multiobjective Genetic Algorithm (moGA) and the Simulated Annealing (SA) are cooperatively used. In this method, the convergence of Genetic Algorithm (GA) is improved by introducing the probability of SA as the criterion for acceptance of new trial solution. The proposed algorithm has multiobjective to minimize the total tardiness and total number of processors used simultaneously. The effectiveness of the proposed algorithm is shown through simulation studies

Keywords: Genetic Algorithm, Multiobjective, Real time Task, Scheduling Algorithm, Simulated Annealing, Total Number of Processors, Total Tardiness

1. Introduction

Real-time systems are classified into two categories: hard real-time system and soft real-time system. In hard real-time system, tardiness can be catastrophic. The goal of hard real-time scheduling algorithms is to meet all tasks' deadlines, in other words, to keep the feasibility of scheduling through admission control. However, in the case of soft real-time systems, slight violence of deadlines is not so critical [1].

In hard real-time system, the performance of scheduling algorithm is measured by its ability to generate a feasible schedule for a set of real-time tasks. Typically, there is Rate Monotonic (RM) and Earliest Deadline First (EDF) derived scheduling algorithms for hard real-time system with uniprocessor [2], [3]. They guarantee the optimality in somewhat restricted environments. However, these algorithms have some drawbacks to cope with soft real-time system related resource utilization and pattern of degradation under the overloaded situation. As the growing of soft real time applications, the necessity of scheduling algorithm for soft real-time system is on increase. Several researches for soft real time system are reported [4, 5]. However, these algorithms also cannot show the graceful degradation of performance under the overloaded situation. Furthermore, the optimal assignment of tasks to multiprocessor is, in almost all practical cases, an NP-hard problem [6]. Consequently various modern heuristics based algorithms have been proposed for practical reason [7].

Recently, several approaches with single objective using Genetic Algorithm (GA) are proposed [8-11]. These algorithms have only one objective such as minimizing cost, end time, total tardiness.

Oh and Wu presented a GA for scheduling nonpreemptive tasks in soft real-time system with multiprocessors [12]. They deal two objectives which are to minimize the total tardiness and total number of processors used. However this algorithm didn't refer about confliction between objectives, so called Pareto Optimum, and has some questions for simulation. In the simulations, deadlines of each task are uniformly generated between from earliest start time to 1.5 times of earliest start time. It means that the earliest start time of latter tasks with long predecessors are much bigger than those of former tasks in task graph. Consequently, along with precedence constraints, the dead lines of latter tasks become dramatically looser one.

In this paper, we propose a new scheduling algorithm for nonpreemptive tasks in soft real-time multiprocessor system with the precedence relationship between tasks. In solution algorithms, the multiobjective Genetic Algorithm (moGA) and the simulated annealing (SA) are cooperatively used [13]. In this method, the convergence of GA is improved by introducing the probability of SA [14] as the criterion for acceptance of new trial solution. However, it is hard to find the optimum solution by only applying the genetic operators.

The objective of proposed scheduling algorithm is to minimize the total tardiness and total number of processors used simultaneously. For these conflicting objectives, this paper combines Adaptive Weight Approach (AWA) that utilizes some useful information from the current population to readjust weights for obtaining a search pressure toward a positive ideal point [15].

2. Soft real-time task scheduling problem and mathematical model

In this study, we consider the problem of scheduling the tasks of precedence and timing constrained task graph on a set of homogeneous processors in a way that simultaneously minimizes the number of processors used f_1 and the total tardiness f_2 under the following conditions:

1. All tasks are nonpreemptive.
2. Every processor processes only one task at a time.

3. Every task is processed on one processor at a time.
4. Only processing requirements are significant; memory, I/O, and other resource requirements are negligible.

The soft real-time task scheduling problem (sr-TSP) is formulated under the following assumptions: Computation time and deadline of each task are known. A time unit is artificial time unit. Soft real-time tasks scheduling problem is formulated as follows:

$$\min f_1 = M \tag{1}$$

$$\min f_2 = \sum_{i=1}^{N} \max\left\{0,\ t_i^S + c_i - d_i\right\} \tag{2}$$

$$\text{s.t.}\quad t_i^E \le t_i^S \le d_i,\quad \forall i \tag{3}$$

$$t_i^E \ge t_j^E + c_j,\quad \tau_j \in \text{pre}(\tau_i),\quad \forall i \tag{4}$$

$$1 \le M \le N \tag{5}$$

In above equations, notations are defined as follows:

- Indices

  i, j: task index, i,j=1,2,…,N
  m: processor index, m=1,2,…,M

- Parameters

  $G=(T, E)$: task graph
  $T=\{\tau_1, \tau_2, \ldots, \tau_N\}$: a set of N tasks
  $E=\{e_{ij}\}$, i,j=1,2,…,N, $i \neq j$: a set of directed edges among the tasks representing precedence
  τ_i: ith task, i=1,2,…, N
  p_m: mth processor, m=1,2,…,M
  c_i: computation time of task τ_i
  d_i: deadline of task τ_i
  pre*(τ_i): set of all predecessors of task τ_i
  suc*(τ_i): set of all successors of task τ_i .
  pre(τ_i): set of immediate predecessors of task τ_i
  suc(τ_i): set of immediate successors of task τ_i .
  t_i^E: earliest start time of ith task

$$t_i^E = \begin{cases} 0, & \text{if } \neg\exists \tau_j : (\tau_j, \tau_i) \in E \\ \max\limits_{\tau_j \in \text{pre}^*(\tau_i)} \{t_j^E + c_j\}, & \text{otherwise} \end{cases},\quad \forall i$$

  t_i^L: latest start time of ith task

$$t_i^L = \begin{cases} d_i - c_i, & \text{if } \neg\exists \tau_j : (\tau_i, \tau_j) \in E \\ \min\{\min\limits_{\tau_j \in \text{suc}^*(\tau_i)} \{t_j^L - c_i\}, d_i - c_i\}, & \text{otherwise} \end{cases},\quad \forall i$$

- Decision Variables
  t_i^S: real start time of ith task
  M: total number of processors used

Equations (1) and (2) are the objective function in this scheduling problem. In equation(1) means to minimize the total number of processors used and equation (2) means to minimize total tardiness of tasks. Constraints conditions are shown from equation (3) to equation (5). Equation (3) means that task can be started after its earliest start time, begin its deadline. Equation (4) defines the earliest start time of task based on precedence constraints. Equation (5) is nonnegative condition for the number of processors.

3. Proposed GA

In this paper, solution algorithm is based on genetic algorithm (GA). Several new techniques are proposed in the encoding and decoding algorithm of genetic string.

3.1. *Encoding and decoding*

A chromosome V_k, $k = 1,2,\ldots,popSize$, represents one of all the possible mappings of all the tasks into the processors. Where *popSize* is the total number of chromosomes in a generation. A chromosome V_k is partitioned into two parts $u(\cdot)$, $v(\cdot)$. $u(\cdot)$ means scheduling order and $v(\cdot)$ means allocation information. The length of each part is the total number of tasks. The scheduling order part should be a topological order with respect to the given task graph that satisfies precedence relationship. The allocation information part denote the processor to which task is allocated.

Encoding procedure for soft real-time task scheduling problem (sr-TSP) is composed of two strategies: strategy I for $u(\cdot)$ and strategy II for $v(\cdot)$. Procedures will be written as follows:

procedure: Encoding Strategy I for sr-TSP
input: task graph data set
output: $u(\cdot)$
begin
    $l \leftarrow 1$, $w \leftarrow \phi$;
    while ($T \neq \phi$)
        $w \leftarrow w \cup \arg\{\tau_i | \text{pre}^*(\tau_i) = \phi, \forall i\}$;
        $T \leftarrow T - \{\tau_i\}, i \in w$;
        while ($w \neq \phi$)
            $j \leftarrow \text{random}(w)$;

```
                    u(l) ← j ;
                    l ← l+1;
                    w ← w − {j };
                    pre*(τ_i) ← pre*(τ_i) − {τ_j}, ∀ i;
                end
            end
            output u(·);
        end
```

procedure: Encoding Strategy II for sr-TSP

input: task graph data set, $u(\cdot)$, $M = \begin{cases} M(k-1), & \text{if } 1 < k \le popSize \\ |\text{subgraph}|, & \text{if } k = 1 \end{cases}$, α, β

output: $v(\cdot)$, M_k

```
begin
    l←1, t_m ← 0, ∀ m, idle ← 0;
    while (l ≤ N)
        m ← random[1, M];
        i ← u(l);
        if (t_m < t_i^E) then
            t_i^S ← t_i^E;
            idle ← idle + (t_i^S - t_m);
        end
        else t_i^S ← t_m;
        if (t_i^S > t_i^L) then
            if (idle/c_i < α) then
                M ← M+1;
                m ← M;
                idle ← idle + t_i^E;
                t_m ← t_i^E + c_i;
            end
            else idle ← max{0, (idle - c_i)} ;
        end
        else
            t_m ← t_i^S + c_i;
            v(l) ← m;
            l ← l+1;
        end
        idle ← idle + ∑(max{ t_m } - t_m);
```

```
        end
        while (idle/∑M x max{t_m}> β)
            M ← M-1;
            idle ← idle- idle/∑M x max{t_m};
        end
        output v(·), M_k;
    end
```

In encoding strategy II procedure, the total number of processors is not fixed. Before generating $v(\cdot)$, the total number of processors M is defined and after generating $v(\cdot)$, M influence to $v(\cdot)$ of next chromosome. The M is changed during generating $v(\cdot)$. So, the M of output is different to the M of input. Where α, β is boundary constant to decide increasing the number of processor and decreasing the number of processor respectively.

Decoding procedure is will be written as follows:

procedure: Decoding for sr-TSP

input: task graph data set, chromosome $u(\cdot)$, $v(\cdot)$

output: schedule set S, the total number of processor used f_1,
total tardiness of tasks f_2

begin

$l \leftarrow 1$, $t_m \leftarrow 0$, $\forall m$, $idle_m \leftarrow \phi$, $\forall m$, $f_1 \leftarrow 0$, $f_2 \leftarrow 0$,
$S \leftarrow \phi$;

while ($l \leq N$)

$i \leftarrow u(l)$;
$m \leftarrow v(l)$;
if (t_m =0) **then** $f_1 \leftarrow f_1 + 1$;
if (exist suitable idle time) **then** insert(i);
start(i);
update_idle();
$f_2 \leftarrow f_2 + \max\{0, (t_i^S + c_i - d_i)\}$;
$S \leftarrow S \cup \{(i, m{:}\ t_i^S - t_i^F)\}$;
$l \leftarrow l+1$;

end

output S, f_1, f_2 ;

end

where insert(i) means to insert τ_i at idle time if τ_i is computable in idle time, start(i) means to assigne τ_i to maximum finish time of all assigned task to p_m, add_idle() means to add idle time to idle time list if idle time is occurred. I^S means the start time of idle duration, I^F means the end time of idle duration, $idle_m$ means the list of idle time and t_m means the maximum finish time of all assigned task to p_m.

The values of two objective f_1 and f_2 are calculated through the procedure of decoding.

3.2. *Evolution function and selection*

The multiobjective optimization problems have been receiving growing interest from researchers with various backgrounds since early 1960. Recently, GAs have been received considerable attention as a novel approach to multiobjective optimization problems, resulting in a fresh body of research and applications known as genetic multiobjective optimizations [16].

In this paper, we combine Adaptive Weight Approach (AWA) [15] that utilizes some useful information from the current population to readjust weights for obtaining a search pressure toward a positive ideal point.

For the examined solutions at each generation, we define two extreme points (maximum: $\boldsymbol{f}^{+}$, minimum: $\boldsymbol{f}^{-}$)

$$\boldsymbol{f}^{+} = \{f_1^{\max}, f_2^{\max}\} \quad (6)$$

$$\boldsymbol{f}^{-} = \{f_1^{\min}, f_2^{\min}\} \quad (7)$$

where $f_q^{\max}$ and $f_q^{\min}$ are the maximal and minimal values for the qth objective as defined by the following equations:

$$f_q^{\max} = \max_k \{f_q(V_k)\},\ q = 1,2 \quad (8)$$

$$f_q^{\min} = \min_k \{f_q(V_k)\},\ q = 1,2 \quad (9)$$

The equation driven above is a hyper plane defined by the following extreme points in current solutions:

$$\begin{matrix}[f_1^{\max} & f_2^{\min}] \\ [f_1^{\min} & f_2^{\max}]\end{matrix} \quad (10)$$

Adaptive moving line defined by the extreme points ($f_1^{\max}$, $f_2^{\min}$) and ($f_1^{\min}$, $f_2^{\max}$) are shown Fig. 1.

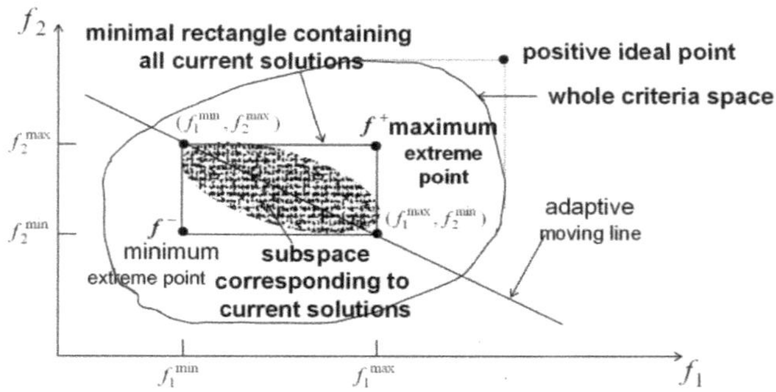

Fig. 1. Adaptive weights and adaptive hyper plane.

The weighted-sum objective function for a given chromosome V_k is given by the following equation:

$$F(V_k) = \sum_{q=1}^{2} w_q f_q(V_k) = \sum_{q=1}^{2} \frac{f_q(V_k)}{f_q^{\max} - f_q^{\min}} \qquad (11)$$

where w_q is adaptive weight for objective q :

$$w_q = \frac{1}{f_q^{\max} - f_q^{\min}}, \quad q = 1,2 \qquad (12)$$

The evaluation function is designed as follows:

$$eval(V_k) = 1/F(V_k) = \frac{1}{\sum_{q=1}^{2} \frac{f_q(V_k)}{f_q^{\max} - f_q^{\min}}} \qquad (13)$$

For selection, the commonly strategy called roulette wheel selection [10], [17] has been used.

3.3. *GA operators*

We use one-cut crossover. This operator creates two new chromosomes (the offspring) by mating two chromosomes (the parent). We crossover only part $v(\cdot)$ in chromosome to maintain the topological order with respect to the given task graph that satisfies precedence relations.

For another GA operator, mutation, we use the classical one-bit altering mutation.

3.4. *Improving of convergence by the probability of SA*

The convergence speed to local optimum of the GA can be improved by adopting the probability of simulated annealing (SA) [11, 12]. Even though the fitness function value of newly produced strings is lower than those of current strings, the newly produced ones are fully accepted in early stages of searching process. However, in later stages, a string with lower fitness function value is seldom accepted. The procedure of improved GA by the probability of SA will be written as follows:

procedure: Improving of GA by the probability of SA
input: parent chromosome V, proto-offspring chromosomes V', temperature T,
    cooling rate of SA ρ
output: offspring chromosomes V''
begin
    $r \leftarrow$ random[0,1];
    $\Delta E \leftarrow eval(V')$-eval($V$);
    if ($\Delta E > 0 \,||\, r < \text{Exp}(\Delta E/T)$) **then** $V'' \leftarrow V'$;
    else
        $V'' \leftarrow V$;
        $T \leftarrow T \times \rho$;
    end
    output offspring chromosomes V''
end

In this procedure, V and V' are mean parent chromosome and proto-offspring chromosome. V'' means offspring chromosome which produced by this procedure. The T means the temperature and the ρ means the cooling rate of SA.

4. Validation

To validate proposed multiobjective hybrid genetic algorithm (mohGA), several numerical tests are performed. We compared proposed mohGA with Oh-Wu's algorithm by Oh and Wu and moGA which is not combined with SA. Numerical tests are performed with randomly generated task graph. We use P-Method [18] for generation task graph. Numerical tests are performed with three task graph: the number of tasks 10, 50 and 100. Task graph and data set is omitted.

Table 1, 2 and 3 shows the comparisons of results by three different scheduling algorithms.

Table 1. Comparison in no-tardiness for 10 tasks.

Terms	Oh-Wu's Algorithm	moGA	mohGA
# of processors *M*	4	3	3
Makespan	29	27	28
CPU times (*msec*)	13	14	19
Average utilization of processors	0.487179	0.550725	0.578571

Table 2. Comparison in no-tardiness for 50 tasks.

Terms	Oh-Wu's Algorithm	moGA	mohGA
# of processors *M*	35	32	26
Makespan	35	37	41
CPU times (*msec*)	118	122	191
Average utilization of processors	0.368339	0.412281	0.573981

Table 3. Comparison in no-tardiness for 100 tasks.

Terms	Oh-Wu's Algorithm	moGA	mohGA
# of processors *M*	36	34	31
Makespan	157	163	168
CPU times (*msec*)	497	511	818
Average utilization of processors	0.417582	0.433392	0.497352

In Tables 1–3, there is no tardiness inclusively. The computing time of proposed mohGA is a little bit longer than those of the other two algorithms. However, the number of used processors is fewer than those of the Oh-Wu's algorithms. The average utilization of processors by mohGA is more desirable than those of the others.

Figures 2, 3 and 4 represent the Pareto solution of proposed mohGA and those of other two algorithms. In these Figures, the Pareto solution curve by proposed mohGA is closer to ideal point than that of other two algorithms.

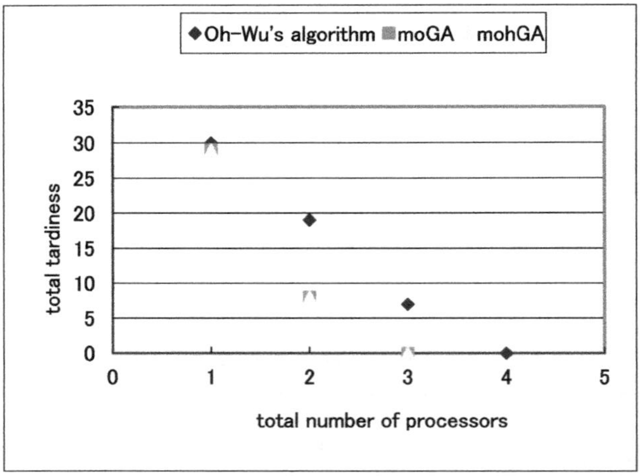

Fig. 2. Pareto solution in 10 task.

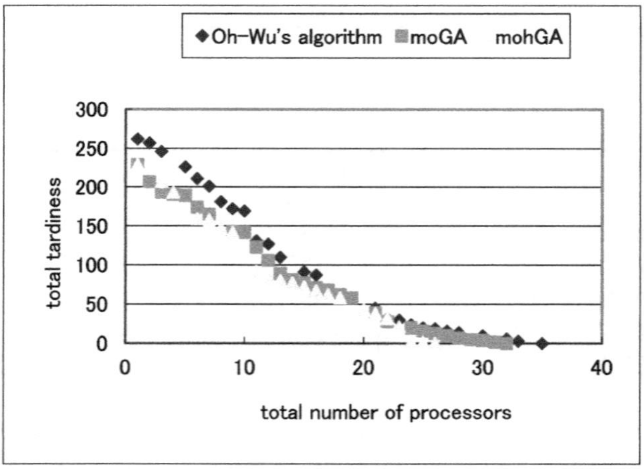

Fig. 3. Pareto solution in 50 tasks.

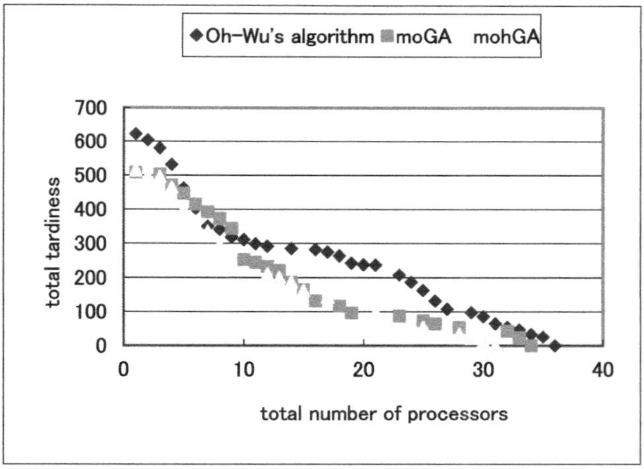

Fig. 4. Pareto solution in 100 tasks.

5. Conclusions

A new task scheduling algorithm is proposed in this paper. This algorithm is designed for nonpreemptive tasks in soft real-time multiprocessor system with the precedence relationship between tasks. In solution algorithms, we proposed a method which combines the multiobjective Genetic Algorithm (moGA) and the simulated annealing (SA). In this method, the convergence of genetic algorithm (GA) is improved by introducing the probability of SA as the criterion for acceptance of new trial solution. The objective of proposed scheduling algorithm is to minimize the total tardiness and total number of processors used simultaneously. For these conflicting objectives, this paper combines Adaptive Weight Approach (AWA). From the numerical results, the results of the proposed mohGA are better than that of other algorithms.

This determines the next step of our study. We plan to design real-time tasks scheduling algorithm in heterogeneous multiprocessors system.

Acknowledgments

This work supported in part by JSPSKAKENHI Grant Number 15K00084.

References

1. C. M. Krishna and G. S. Kang, *Real-Time System* (McGraw-Hill, 1997).
2. J. L. Diaz, D. F. Garcia and J. M. Lopez, Minimum and Maximum Utilization Bounds for Multiprocessor Rate Monotonic Scheduling, *IEEE Transactions on Parallel and Distributed Systems* **15**, 7 (2004) pp. 642-653

3. G. Bernat, A. Burns and A. Liamosi, Weakly Hard Real-Time Systems, *Transactions on Computer Systems*, **50**, 4 (2001) pp. 308-321.
4. M. H. Kim, H. G. Lee and J. W. Lee, A Proportional-Share Scheduler for Multimedia Applications, *Proc. of Multimedia Computing and Systems* (1997) pp. 484-491.
5. M. R. Yoo, A Scheduling Algorithm for Multimedia Process, Ph.D. dissertation, University of YeoungNam, (Korea, 2002).
6. J. M. Landsberg, L. Manivel, F. Yalaoui and C. Chu, Parallel Machine Scheduling to Minimize Total Tardiness, *International Journal of Production Economics*, **76**, 3 (2002) pp. 265-279.
7. M. Yoo and T. Yokoyama, Multiobjective GA for Real Time Task Scheduling, *Lecture Notes in Engineering and Computer Science: Proceedings of The International MultiConference of Engineers and Computer Scientists 2016, IMECS 2016*,16-18 March, 2016, Hong Kong, pp. 174-178.
8. H. Mitra and P. Ramanathan, A Genetic Approach for Scheduling Non-preemptive Tasks with Precedence and Deadline Constraints, *Proc. of the 26th Hawaii International Conference on System Sciences* (1993) pp. 556-564.
9. M. Lin and L. Yang, Hybrid Genetic Algorithms for Scheduling Partially Ordered Tasks in A Multi-processor Environment, *Proc. of the 6th International Conference on Real-Time Computer Systems and Applications* (1999) pp. 382-387.
10. Y. Monnier, J. P. Beauvais and A. M. Deplanche, A Genetic Algorithm for Scheduling Tasks in a Real-Time Distributed System, *Proc. of 24th Euromicro Conference* (1998) pp. 708-714.
11. M. Yoo, Continuous Media Tasks Scheduling Algorithm, *International Journal of Electronics Communication and Computer Engineering*, **7**, 2 (2016) pp. 99-103.
12. J. Oh and C. Wu, Genetic-algorithm-based Real-time Task Scheduling with Multiple Goals, *Journal of Systems and Software*, **71**, 3 (2004) pp. 245-258.
13. H. Kim, Y. Hayashi and K. Nara, An Algorithm for Thermal Unit Maintenance Scheduling through combined use of GA SA and TS, *IEEE Trans. on Power Systems*, **12**, 1 (1997) pp. 329-335.
14. S. Kirkpatrick, C. D. Gelatt and M. P. Vecchi, Optimization by Simulated Annealing, *Science*, **220**, 4598 (1983) pp. 671-680.
15. M. Gen and R. Cheng, *Genetic Algorithms & Engineering Optimization* (John Wiley & Sons, 2000).

16. K. Deb, *Multi-objective Optimization using Evolutionary Algorithms* (John Wiley & Sons, 2001).
17. M. Gen and R. Cheng, *Genetic Algorithms & Engineering Design* (John Wiley & Sons, 1997).
18. S. Al-Sharaeh and B. E. Wells, A Comparison of Heuristics for List Schedules using The Box-method and P-method for Random Digraph Generation, *Proc. of the 28th Southeastern Symposium on System Theory* (1996) pp. 467-471.

A COMPARATIVE THEORETICAL AND BENCHMARKING EVALUATION OF MODERN OPERATING SYSTEM KERNELS

STERGIOS PAPADIMITRIOU* and LEFTERIS MOUSSIADES†

Computer and Informatics Engineering Department Eastern Macedonia & Thrace Institute of Technology of Technology
Agios Loukas, 65404 Kavala, Greece
**sterg@teiemt.gr*
†lmous@teiemt.gr

The paper compares core kernel architecture and functionality of four modern operating systems. The subsystems examined are process/thread architecture, scheduling and interrupt handling. Linux, Solaris and FreeBSD have a lot of similarities, owning UNIX roots, but also have some notable differences. However, Windows is significantly different, being a radical non-UNIX design. The paper compares some aspects of the UNIX-like approaches of Linux/Solaris/FreeBSD with Windows, emphasising the consequences of their different design decisions, and presents comparative benchmarking performance results.

Keywords: interrupts, process, scheduling, threads

1. Introduction

In this work we examine comparatively core subsystems of three modern UNIX-like operating systems and contrast them with the radical non-UNIX approaches of Windows. The subsystems examined are *process* and *thread* structure, *interrupt structure* and *scheduling*.

It is an extension over the work presented in Ref. 1. Specifically, the UNIX based systems compared are the Linux, Solaris and FreeBSD operating systems. The current 4.x Linux kernel incorporates many advanced features[5–7] and stands well compared to the also state of the art Solaris 11 kernel[3,4,8] and the FreeBSD 11 kernel.[10] Also, some aspects of modern Windows (Windows 7,8,10)[9] are compared with UNIX designs.

The paper concludes that Linux, FreeBSD and Solaris systems are modern and effective UNIX realizations, that their UNIX philosophy offers to them many similarities.[2] Linux seems the most efficient, both from the point of view of the user experience (which is subjective) and from actual performance results. This efficiency can be explained from the fact that Linux generally uses less intermediate layers of code before accessing hardware. Windows implements different design solutions, and it is interested to study them and explore their benefits and drawbacks.

The paper proceeds as follows: Section 2 highlights some important aspects of the basic architecture of these systems. Section 3 exploits the interrupt processing architecture that is of outstanding importance for the efficient support of higher layers and for meeting real-time processing demands. Section 4 studies the scheduling approaches of the four systems and compares them. Section 5 highlights some basic design differences between Windows and UNIX-like systems. Section 6 describes the benchmarking framework and discusses on the results. The results are obtained using the *pthreads* standard C library instead of the Java benchmarks presented in.[1] Finally, the paper concludes with the results of this comparative study.

2. Basic Architecture

Both Linux, Solaris, FreeBSD and Windows are monolithic kernels. This organization permits efficient execution since kernel components interact with direct procedure calls without involved with message passing and the associated switches between user and kernel mode code. Windows started as a microkernel design,[9] but performance demands forced the movement of components designed to operate in user mode (for benefit of reliability and modularity) again in kernel space. The recent Windows kernel has a modular design with distinct kernel components in different modules. However, all the modules operate in kernel space and therefore routines of one module can directly call routines from other modules. This way, Windows kernel avoids a great deal of overhead. The modular but yet monolithic design of Windows kernel is referred sometimes as *macrokernel* design.[9]

The basic unit of scheduling in Solaris is the *kthread_t* structure;[3] and in Linux, the *task_struct* structure.[5] Solaris represents each process as a *proc_t*, and each thread within the process has a *kthread_t*. Linux represents both processes and threads by *task_struct* structures. A single-threaded process in Linux has a single *task_struct*. A single-threaded process in Solaris has a *proc_t*, a single *kthread_t*, and a *klwp_t* structure. The *klwp_t* structure provides kernel state for user threads, i.e. is a save area for threads switching between user and kernel modes. Effectively, all operating systems schedule threads. In Linux a thread corresponds to a *task_struct* structure and in Solaris a thread is a *kthread_t*. Multithreaded processes in Linux are implemented as normal processes (i.e. *task_struct* structures) that share the same address space and resources.

FreeBSD like Solaris implements a multithreaded process design.[10] Each FreeBSD process (structure *proc*) keeps a linked list of its threads(structure *thread*). These threads are scheduled by the kernel, and they own their own kernel stacks onto which multiple threads of the same process can execute system calls simultaneously. The thread structure of FreeBSD represents just the information needed to run in the kernel: information about scheduling, a stack to use when running in the kernel, a *thread state block*, and other machine dependent state.

FreeBSD organises sleep and turnstile queues in a data structure that is hashed by an event identifier. Turnstile queues implement an elaborate scheme that supports priority inversion. The priority inversion support along with the possibility of preempting the kernel when real-time and interrupt threads exist, constitute FreeBSD capable of supporting real-time applications. FreeBSD utilizes bottom-half interrupt priorities and top-half interrupt priorities and maps them to corresponding threads. Bottom-half interrupt threads are the more critical, next come the real-time and the top-half interrupt threads follow. For each thread a user mode execution priority and a kernel mode execution priority is assigned. Threads sleeping in the kernel have a higher priority, because they typically hold shared kernel resources when they awakened. Also, FreeBSD provides restartable system calls.

Windows implements threads distinctly, as Solaris and FreeBSD do, and not as Linux that views threads as normal processes with shared resources. Windows processes are represented by an *executive process* (*EPROCESS*) block. Threads are represented by *executive thread* (*ETHREAD*) blocks. The EPROCESS is a multithreaded process (like Solaris *proc_t*), therefore it links to one or more ETHREAD (like Solaris *kthread_t*) blocks. The part of the EPROCESS structure that needs to be accessed from user-mode code, is isolated to the *Process Environment Block* (*PEB*) (like Solaris *klwp_t*) that exists in the process address space. Thus, PEB in Windows plays a similar role as *klwp_t* of Solaris, i.e. a glue for switching user and kernel modes of a process.

3. Interrupt Structure

Although interrupt controllers perform a level of interrupt prioritization, Windows imposes its own interrupt priority scheme, on top of the interrupt prioritization performed by interrupt controllers, known as *interrupt request levels (IRQLs)*. The IRQLs aren't the same as the interrupt requests (IRQs) defined by interrupt controllers. The kernel represents IRQLs internally as a number from 0 through 31 on x86 and from 0 to 15 on x64 and IA64. Higher numbers represent higher priority interrupts. Although the kernel defines the standard set of IRQLs for software interrupts, the Hardware Abstraction Layer (HAL) maps hardware interrupt numbers to the IRQLs.

Windows uses a type of device driver called a *bus driver*, to determine the presence of devices on a bus type (e.g. PCI, USB etc.) and what interrupts can be assigned to a device. The bus driver reports this information to the *Plug and Play manager* which decides, after taking into account the acceptable interrupt assignments for all other devices, which interrupt will be assigned to each device. Then it calls a *Plug and Play interrupt arbiter*, which maps interrupts to IRQLs.

Both Windows and Linux use a system table called *Interrupt Descriptor* (*or Dispatch*) *Table* (*IDT*). The IDT associates each interrupt or exception vector with the address of the corresponding interrupt or exception handler. The IDT must be properly initialized before the kernel enables interrupts. At system boot the IDT is filled with pointers to the kernel routines that handle each interrupt and exception. Each processor in Windows has a separate IDT, so that different processors can run different ISRs, if appropriate. Windows statically assigns interrupts to processors in a round-robin manner.

Windows and Linux use the kernel stack of the current process for interrupt frames and therefore they cannot reschedule from an interrupt handler. This is because interrupts can be arbitrarily nested, stacking multiple contexts on the kernel stack. If we reschedule, the stack with the saved interrupt frames is lost, and there is a problem at restoring those contexts after the interrupt handler finishes. uses the kernel stack of the currently executing thread for servicing interrupts. Linux does the same when two-page stacks (i.e. 8Kb) are used. However, for one page stack (i.e. 4 Kb) Linux uses separate *exception*, *hard IRQ stack* and *soft IRQ* stacks. These are multiple kernel mode stacks: for interrupt handling, for exceptions (one per-process), for hard IRQ (one per CPU) and for soft IRQs (one per CPU).

Making a difference from Windows, Linux implements an interrupt model without priority levels. Each interrupt handler may be deferred by another one, that corresponds to a higher priority interrupt line. Therefore,there is no need to establish additional predefined priorities among hardware devices beyond those defined by the hardware interrupt arbiter. This simplifies the kernel code and improves its portability.

Linux implements an elaborate scheme to balance the interrupt load among the processors at a multiprocessor environment. Specifically, a special kernel thread, the *kirqd*, exploits the IRQ affinity of a CPU by modifying the *Interrupt Redirection Table* entries of the I/O APIC. The later table controls the distribution of interrupt processing among CPUs. Therefore, Linux obtains generally a balanced load of the interrupt workload over multiprocessors.

Linux provides three different and complementary ways to perform deferred processing: *Softirqs* , *Tasklets* and *Work queues*

Softirqs is the most efficient mechanism. They are statically allocated (i.e. defined at compile time). They can run concurrently on several CPUs, even if they are of the same type. Therefore, to fullfill these concurrency requirements, they must be *reentrant* functions and must explicitly protect their data structures with spin locks.

Tasklets are built upon the softirq mechanism but they can be dynamically allocated (i.e. at runtime). That makes them proper for use with dynamically loaded kernel modules. Contrary, to softirqs, they can be *non-reentrant* functions. Therefore, they are easier to write, but they are imposed to serialization, i.e. tasklets of the same type cannot run in parallel on several CPUs. However, tasklets of different types can.

Both softirqs and tasklets execute in *interrupt context*, therefore they cannot sleep. This imposes significant limitation on what the programmer can perform within a softirq-tasklet.

The *work queue* mechanism is introduced to overcome this limitation. Work queues schedule deferrable work within the process context of special kernel threads. Thus this deferable code, is schedulable, and can therefore sleep. However, work queues induce the overhead of scheduling and of the necessary context switches. Softirqs and tasklets are a more efficient mechanism since they are executed totally within kernel space, as simple C routine calls. We note that although work queues imply a context switch, that context switch is efficient, since switching takes place totally within kernel space also. Practically, work queues are quite efficient, but when we have to service thousands of interrupts per second, using tasklets or even the more efficient softirqs makes sense.

Linux also supports an elaborate distribution of IRQ signals coming from the hardware devices in a round-robin fashion among all the CPUs.[5] The Linux approach to interrupt handling stays closer to hardware from the other OSes. A competent C programmer can readily write interrupt handlers without having to familiarize with complicated layering e.g. the "interrupt objects" of Windows. Linux implements an interrupt architecture that although rather low-level, it exploits fully the parallelism on an SMP system and is simple to use.

Solaris uses the notion of the *interrupt thread* for servicing interrupts. This has similarities to the *work queues* of Linux. Upon recept of the interrupt, the current thread is pinned and the interrupt thread executes. When the interrupt thread completes, the interrupted thread is unpinned and resumes execution. This mechanism avoids a full context switch for servicing interrupts. Solaris interrupt threads can sleep also: If the interrupt thread blocks, it is given full thread state and it is placed on a sleep queue, and the interrupted thread will be unpinned. Thus, for interrupt code that doesn't sleep, Solaris interrupt threads can be more efficient than Linux work queues.

Solaris supports the *cross-call mechanism* for multiprocessor systems. It is a facility with which one CPU can send an interrupt to one or more other CPUs on the system (or to all of them) to force the CPU into a handler to take a specific action.

Solaris (as Windows also) assigns priorities to interrupts to allow overlapping interrupts to be handled with the correct precedence. This interrupt prioritization performs similarly to the IRQL based prioritization performed by the Windows kernel. We should note that Linux does not prioritize interrupts explicitly, but by allowing nesting of interrupts enforces the interrupt prioritization that the interrupt controller implements. However, interrupt handlers using priority levels alone cannot block, since a deadlock could occur if they are waiting on a resource held by a lower priority interrupt.

FreeBSD uses also high-priority *interrupt threads* as Solaris does. The highest priority threads of class *ITHD* serve the time critical demands for interrupt processing, tasks that on single processor systems were usually performed within the interrupt service routine by disabling the CPU interrupts.

Scheduling decisions are based on *priority*. In Linux, the lower the priority value, the better, i.e. a value closer to 0 represents a higher priority. In Solaris, the higher the value, the higher the priority. For Linux, the priority range 0–99 corresponds to the *System Threads, Real-Time* Scheduling Class (*SCHED_FIFO, SCHED_RR*) and the priority range 100–139 to *User priorities* (*SCHED_NORMAL*). In Solaris, the priority range 0–59 corresponds to the *Time Shared, Interactive, Fixed, Fair Share Scheduler* class, the range 60–99 to the *System Class*, the range 100–159 to the *Real-Time* and finally the highest priority range 160–169 to the *Low level Interrupts*.[3,5] Both Windows, FreeBSD, Solaris and Linux use a per-CPU *runqueue*.

FreeBSD's highest priority *s* deserved for interrupt threads of class ITHD. Following the ITHD scheduling class, comes in priority the *REALTIME* class. Therefore, with the prerequisite of small and bounded overhead for the ITHD thread processing, the engineer can design and implement on FreeBSD real time processing workloads, The *KERN* class follows the REALTIME class in priority. This class performs deferred interrupt processing, i.e. is the top-half kernel processing of the interrupt requests. Time consuming parts of the interrupt service tasks should be implemented within the threads of the KERN class. The class that runs the "normal" user applications, the *TIMESHARE* follows in priority. The kernel adapts dynamically the priority of the threads of this class in order to provide better response to the interactive tasks. These priorities are adjusted based on resource usage and recent CPU utilization. A thread has two scheduling priorities: one for scheduling user-mode execution and one for scheduling kernel mode execution. Finally, the *IDLE* class consumes the CPU time, when no useful task exists.

All OSes favor interactive threads/processes. Interactive threads run at better priority than compute-bound threads, but tend to run for shorter time slices. Traditionally, time-slice based scheduling uses heuristic criteria to favor interactive tasks. This can be suboptimal and problematic.

Linux introduces a new type of scheduling the *Completely Fair Scheduling* that avoids the problem of computing the proper time slices[7] The Completely Fair Scheduler (CFS) tries to approximate the concept of perfect multitasking, i.e. the fair sharing of processor time by the tasks. Suppose, for example, that we have N tasks of equal priority to schedule. The CFS should try to approximate the ideal situation of having them running as if each task was alone at the CPU, of course running at $\frac{1}{N}$ speed.

Implementing such an ideal protocol requires to have the processes run for infinitely small durations. However, in practice is not efficient to run the tasks for very small durations, due to context switching overheads and the effect of switching on caches. The larger are the time slices (i.e. time quantums) the smaller is the overall overhead, and the better the overall system throughput, at the expense of increasing task latencies. CFS intelligently settles this tradeoff by fixing the *targeted latency*, i.e. the maximum affordable latency for a set of tasks. Then CFS computes, based on the latency constraint, a schedule that shares the processor time fairly and optimizes switching overhead, i.e. it performs only the necessary task swaps to maintain the latency constraints. Clearly, the larger the task workload, the smallest the computed timeslice to achieve the specified targeted latency.

CFS uses weighting factors to implement different task importances, i.e. priorities. The CFS also avoids starving low priority tasks. With CFS irrespective of how much heavily loaded is a system, even the less important tasks receive some time slice.

The switching rate for tasks with the timeslicing based Windows, FreeBSD and Solaris scheduling is dependent on the load and cannot fullfill systematically the required latencies. In contrast, the Completely Fair Scheduler (CFS) of Linux, designs the scheduling starting with the constraint of maintaining the required latencies.

Quantum based scheduling systems usually perform a priority boost after a release from a wait operation. Since usually the interactive I/O bound tasks are blocked, this strategy favors them. Also, they adaptively decrease the priority of compute bound tasks on every successive time quantum that they take. The Completely Fair Scheduler automatically favors I/O bound tasks, since their blocking time provides them an advantage to take their turn. The traditional priority boost mechanism is rather heuristic and it cannot guarantee that interactive tasks get timely the CPU, as the target latency based timeslice computation of CFS does.

Solaris uses a *dispatch queue* per CPU. If a thread uses up its time slice, the kernel gives it a new priority and returns it to the dispatch queue.

The *runqueues* for all operating systems have separate linked lists of runnable threads for different priority ranges. Both Solaris and Linux use a *separate list* for each priority. FreeBSD groups priorities in priority ranges and uses a list per priority range (in order to use fewer lists). The previous O(1) scheduler of Linux used an arithmetic calculation based on run time versus sleep time of a thread (as a measure of *interactiveness*) to arrive at a priority for the thread. Similar style computations are performed by the ULE scheduler of FreeBSD and the Solaris and Windows schedulers.

All OSes schedule the one next thread to run, instead of attempting to derive a schedule for a whole group of n threads. Also, they have mechanisms to take advantage of caching (warm affinity) and load balancing. For hyperthreaded CPUs, Solaris has a mechanism to help keep threads on the same CPU node. This mechanism is under control of the user and application.

One of the big differences between Solaris and Linux is the capability of Solaris to support multiple *scheduling classes* on the system at the same time. The ability to add new scheduling classes to the system comes with a price. Everywhere in the kernel that a scheduling decision can be made (except for the actual act of choosing the thread to run) involves an indirect function call into scheduling class-specific code. For instance, when a thread is going to sleep, it calls scheduling-class-dependent code that does whatever is necessary for sleeping in the class. On Linux the scheduling code simply does the needed action. There is no need for an indirect call. The extra layer means there is slightly more overhead for scheduling on Solaris, but more supported features. Similar to Linux, FreeBSD avoids the dynamic selection of the scheduling policy for performance reasons. However, it allows the selection of the scheduler when the kernel is built.[10]

5. Basic Aspects of UNIX-Windows Compared

This section discusses some basic differences between a UNIX-like OS and Windows. We refer mostly to Linux as an example of a UNIX system, although Solaris and FreeBSD fit also well to much of the discussion.

UNIX roots the system name space in the file system. Windows instead uses the *object manager*[9] that constructs an abstract name space and connects the file systems as devices. The object manager also supports directory and symbolic links in the name space. For the object manager, a name can be either permanent or temporary. Permanent names are used for entities such as disk drives, that remain no matter if processes are accessing them. A temporary name is created while a process holds a handle to an object. However, it is not limited to files: processes, threads, synchronization events, semaphores, mutexes are some other entities that the object manager handles.

The Windows Object Manager performs integrated object handling. A distant analog for Linux is the *slab layer*. (Solaris and FreeBSD also use a slab layer). The later performs such memory structure allocation chores (i.e. "Object" allocation) but at a significantly lower layer. A disadvantage of the centralized object handling with the Windows Object Manager is the consequent coupling of modules since all operations go through the object manager, e.g. to implement a file system we need to fully understand the Object Manager. As an example for understanding the coupling between modules that is involved, is that for a file open operation, both the Object Manager, the I/O manager and the file system are involved.

Process creation in Windows involves interaction between the environment subsystem, the object manager and the kernel. Therefore, it is more complicated than corresponding Linux, FreeBSD and Solaris approaches.

Windows implements the *Hardware Abstraction Layer* (*HAL*) that exports a virtual machine interface to the hardware. That interface is used by the kernel disparcher, the executive and the device drivers. The device drivers still have to cope with the hardware details of their specific devices. The HAL however abstracts the details of the particular motherboard. The Linux approach is to present standard interfaces for low-level functions. These interfaces are implemented (of course) by machine dependent source code files that reside at architecture dependent directories of the Linux source tree. This layer of machine dependent code captures the low level details and presents a uniform interface to upper layers. Therefore, it acts as the Windows HAL module. FreeBSD and Solaris isolate similarly the low-level machine code. FreeBSD also uses the CAM (Common Access Methods) layer for device independent disk I/O.

The Windows API functions are documented, callable subroutines. Examples include *CreateProcess*, *CreateFile*, and *GetMessage*. Their functionality is similar to the routines of the Linux *glibc* library. Windows user applications don't call Windows system services directly, but instead they go through one or more *subsystem DLLs*. Many calls of the subsystem DLLs also need to interface system services exported by the Windows Executive, with the *Ntdll.dll* library. These libraries export the documented interface that the programs linked to their subsystem can call. The Windows Subsystem DLLs provide similar functionality to the glibc of Linux, but they are more indirect, since they perform calls from the documented to the undocumented interface.

The Linux kernel structures can be directly accessible from drivers. In Windows, to successfully make use of an object, device drivers need to know the internal structure definition of the object, and this is not provided for most objects. Instead, device drivers are encouraged to use the appropriate kernel APIs to modify or read information from the object. For example, although device drivers can get a pointer to the Executive Process object (*EPROCESS*), the structure is opaque, and *Ps* API* must be used.

Each Windows thread is actually two threads: a *user-mode* thread (UT) and a kernel mode thread (KT). Each has its own stack, register values and execution context. System calls are initiated at the UT. The UT in turn causes a trap to kernel mode. The kernel runs a trap handler that switches between the UT and the corresponding KT. Finally, when the KT finishes, it switches again to the UT. As we discussed, Linux does not separate threads from processes: threads are processes with shared resources. Therefore, to handle a system call a Linux process simply switches context to kernel space. The kernel handles the system call on behalf of the specific process.

For Linux, all the core operating system services are part of a single module, the *vmlinuz* and for Windows the corresponding module is *ntoskrnl.exe* Windowing is handled differently: Windows has a kernel-mode windowing subsystem Linux has a user-mode X-Window based windowing.

Windows drivers provide support for modularity: they allow for dynamic extension of kernel functionality, in essence drivers are dynamically loadable kernel modules. Significant amount of Windows code run as drivers (including network stacks such as TCP/IP and many services) They are built independently from the kernel can be loaded on-demand. Dependencies among drivers can be specified.

For Linux, a *device driver* is the set of kernel routines that makes a hardware device respond to the programming interface defined by the canonical set of VFS functions (open, read, lseek, ioctl, etc) that control the device. Thus, the Linux device driver model is tightly and elegantly integrated with the whole VFS design.

Trap handlers in Windows are similar to Linux. In both systems, the trap refers to a processor's mechanism for capturing an executing thread when an exception or an interrupt occurs and transferring control to a fixed location in the operating system. In both systems, the processor transfers control to a trap handler, a function specific to a particular interrupt or exception. System calls are implemented with exceptions both in Windows and Linux.

Windows *Deferred Procedure Calls* (*DPC*) are like the *tasklets* of Linux. They implement a similar bottom-halve mechanism. Windows DPCs as Linux tasklets cannot sleep. Also, Windows provides *worker threads* that can be used to perform the bottom have chores in process context. Worker threads therefore correspond to the *work queues* of Linux.

6. Benchmarks

The presented benchmarks access the performance of the OSes on a variety of workloads using C low level based code based on the pthreads library. We access also performance using the cygwin (https://www.cygwin.com/) UNIX compatible library. For better representation of the UNIX platforms we also present Mac OS benchmarks. https://www.cygwin.com/ platforms that we used to run the benchmarks are an iMac and a HP PC of about equivalent hardware. We used the PC in order to be able to compare with Linux performance, since we had problems installing Linux on the iMac. Both the used iMac and HP PC have an 3200MHz Intel i5 processor with memory size 8192 Mb.

The set of benchmarks that we evaluated are from the Byte magazine.[11] The results are displayed with Table 1.

The *execl* (i.e. execl 20) column of Table 1 displays the average count of execl calls over 20 secs time, i.e. the effectiveness of creating a process from an executable is measured. Windows is about five times slower than FreeBSD. The process that executed with *exec* is the executable file of the execl benchmark program itself.

The *syscall* benchmark the overhead of performing system calls. Here, Mac OS is slightly faster than FreeBSD, Linux outperforms both, being about two times faster and Windows have a rather poor performance. The syscall benchmark performs sequentially the following system calls: *close*(), *dup*(), *getpid*(), *getuid*(), *umask*().

The *spawn* benchmark exercises the *fork*() system call, i.e. it measures the efficiency of creating processes. Mac OS can be considered to have a slow fork() system call, with Linux being the most efficient, FreeBSD a close competent and Windows is the slowest.

The *context* benchmark creates two pipes p_1 and p_2. It then proceeds by forking a child process. The parent process closes the input descriptor of p_1 i.e. $p_1[0]$) and the output descriptor of p_2 (i.e. $p_2[1]$). The child process does the reverse. Then the two processes communicate using these pipes: the information from parent to child is transferred using the p_1 pipe and the p_2 pipe transfers from the child to the parent. FreeBSD displays the best performance.

The *hanoi* is a purely computational benchmark on which all UNIXes display similar performance (actually FreeBSD is slightly faster) and Windows perform better.

The *pipe* is the simplest form of communication between processes. *Pipe throughput* is the number of times (per second) a process can write 512 bytes to a pipe and read them back. Clearly, FreeBSD performs more efficiently than Mac OS and Windows is much slower.

The *fstime* benchmarks measure file system related performance aspects. FreeBSD demonstrates rather poor performance at the *fstime* benchmarks for write operations, but this is due to the synchronous writes of directory operations in order to maintain consistency.

The *simple read*, *simple write*, *simple open/close*, *simple stat*, *simple fstat*, measure the efficiency of the corresponding operations. Here, Mac OS and FreeBSD perform similarly but Linux is clearly more efficient.

Table 1. Byte magazine Unix benchmarks results.

Benchmark	Windows 8	Mac OS	FreeBSD iMac	FreeBSD PC	Linux PC
execl	5418	7139	24110	33347	49469
syscall	2861812	12884110	10620877	14665300	28773938
spawn	10965	35638	71768	120943	168546
context switch	2861812	1597858	2484023	3261091	2312248
hanoi	7412573	4603145	5025719	4575185	4364714
pipe	7161783	25591689	44677193	39667978	45399029
fstime write	783641	1686255	102547	187417	1211860
fstime read	944584	1862444	1752057	512512	4140431
fstime copy	376006	812984	96719	126403	862759

7. Conclusions

The paper aimed to provide some insight to some key design approaches and components of Linux, Solaris and Windows using a comparative view. Both Windows, Linux and Solaris are monolithic kernels (although Windows started as microkernel). Also, they allow the kernel to be extended flexibly with dynamically loaded modules. Windows imposes more layers between the hardware and the user space applications (e.g. the HAL, the interrupt objects). Microsoft virtualizes part of the hardware and builds other components upon that virtualization. Clearly, this is an advantage from the point of view of portability to different architectures but requires learning additional Microsoft APIs and being dependent on them. From the three OSes, Linux is generally the closest to the hardware, and combined with the clever and highly optimized implementation approaches, realize the fastest overall performance.

Solaris uses more data abstraction layering, and generally could support additional features quite easily because of this. However, most of the layering in the kernel is undocumented.

The application level interface to these systems is very similar, typical of the modern UNIX system programming interface.[8] The paper concludes that both systems are modern and effective UNIX realizations. They are capable of accomplishing effectively demanding application requirements and at the same time their UNIX philosophy offers to them many similarities.[2]

References

1. Stergios Papadimitriou and Lefteris Moussiades, "A comparative Evaluation of Core Kernel Features of the Recent Linux, FreeBSD, Solaris and Windows Operating Systems", Lecture Notes in Engineering and Computer Science: Proceedings of The World Congress on Engineering and Computer Science 2016, London, U.K. 29 June–1 July, 2016, pp. 387-393
2. W. Richard Stevens and Stephen A. Rago, *Adavanced Programming in the UNIX Environment*, 3rd Edition, Addison-Wesley, 2013
3. Richard McDougall and Jim Mauro, *Solaris(TM) Internals: Solaris 10 and OpenSolaris Kernel Architecture*, 2nd Edition (Solaris Series), Sun Microsystems Press, 2006
4. Richard McDougall, Jim Mauro and Brendan Gregg, *Solaris Performance and Tools*, Sun Microsystems Press, 2006
5. Daniel Plerre Bovet and Marco Cesati, *Understanding the Linux Kernel*, O' Reilly, 2005
6. Jonathan Corbet, Alessandro Rubini and Creg Kroah-Hartman, *LINUX Device Drivers*, 3rd Edition, O'Reilly 2005
7. *Linux Kernel Development* (3rd Edition) (Novell Press) by Robert Love, 2010
8. Rich Teer, *Solaris System Programming*, Addison-Wesley, 2005
9. Mark E. Russinovich, David A. Solomon and Alex Ionescu, *Windows Internals*, 6th edition, 2012, Microsoft Press, Vol. I and II
10. Marshall Kirk McKusick, George V. Neville-Neil and Robert N.M. Watson, *The Design and Implementation of the FreeBSD Operating System*, Addison-Wesley, 2014
11. UNIX Byte benchmarks, https://github.com/kdlucas/byte-unixbench

ENHANCED EFFICIENCY OF LARGE WIND POWER GENERATOR APPLICATIONS USING POWER FACTOR CONTROLLER MECHANISM

J. K. OBICHERE* and M. G. JOVANOVIC†

Department of Physics and Electrical Engineering, Northumbria University, Newcastle upon Tyne, NE1 8ST, UK
**judekennedyobichere@yahoo.com,*
†milutin.jovanovic@northumbria.ac.uk
www.northumbria.ac.uk

This work presents a vector control (VC) of an emerging brushless doubly-fed reluctance (BDFRG) technology for large wind turbine applications. The BDFRG has been receiving increasing attention due to its low operation and maintenance costs afforded by the use of partially-rated power electronics, and the high reliability of brushless assembly, while offering performance competitive to its traditional slip-ring counterpart, the doubly-fed induction generator (DFIG). A robust VC strategy has been developed for a custom-designed BDFRG fed from a conventional 'back-to-back' IGBT converter. Preliminary studies have evaluated the algorithm under the power factor control (PFC) conditions which allow the enhanced efficiency of the generator-converter set and the entire wind energy conversion systems (WECS).

Keywords: Brushless; Doubly-Fed Generators; Enhanced Efficiency; Power Factor Control; Vector Control; Wind Energy.

1. Introduction

The brushless doubly-fed generator (BDFG) has been considered as a viable replacement to the traditional DFIG for wind turbines [1]. In these applications, where only a limited variable speed capability is required (e.g. typically, in a 2:1 range or so [1, 4]), the BDFG should retain the DFIG economic benefits of using a relatively smaller inverter (e.g. around 25% of the machine rating), but with higher reliability and maintenance-free operation afforded by the absence of brush gear [5].

The BDFRG has two standard stator windings of different applied frequencies and pole numbers, unlike the DFIG. The primary (power) winding is grid-connected, and the secondary (control) winding is normally supplied from a bi-directional power converter. A BDFG reluctance type (Fig. 1), the brushless doubly-fed reluctance generator (BDFRG), appears to be more attractive than its 'nested' cage rotor form, the brushless doubly-fed induction generator (BDFIG) [6, 7]. This preference has been mainly attributed to the prospect for higher efficiency [2] and simpler control associated with the cage-less reluctance rotor [8]. However, the BDFG rotor must have half the total number of stator poles to provide the rotor position dependent magnetic coupling between the stator windings required for the machine torque production.

With the recent introduction of the grid codes [9], another important BDFG merit is the superior low-voltage-fault-ride-through (LVFRT) capability to the DFIG [10–12]. It has been shown that owing to the larger leakage inductances and lower fault current levels, the LVFRT of the BDFIG may be accomplished safely without a crowbar circuitry [9, 13]. These potential LVFRT performance advantages over the DFIG can be carried over to the BDFRG featuring the leakage reactance values of the same order as the BDFIG.

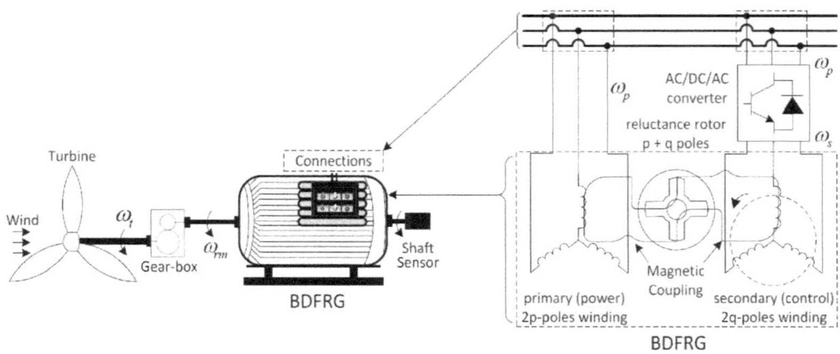

Fig. 1. Typical BDFRG drive set-up of variable speed WECS.

Various control algorithms have been developed for the BDFRG including scalar control [1], vector control (VC) [1, 17], direct torque control, torque and reactive power control [14], direct power control, sliding mode power control, field-oriented control [15], and even non-linear Lyapunov control theory. Although a comparative analysis of some of these control methods has been partly made in [15] (and more detailed for the DFIG in [10]), it is interesting that there is very little reported specifically on Power Factor Control (PFC) of the BDFRG being of utmost importance for generator applications [13]. In the BDFRG, the 'VC' term is commonly referred to as the primary winding voltage control, by analogy to the stator voltage control of the DFIG.

With a proper selection of the reference frames and careful tuning of the dedicated PI controllers, a very good potential and dynamic response of the VC scheme has been demonstrated without knowledge of any machine parameters. The VC performance is examined using the power factor control (PFC) mechanism [14, 18–21] on a large, custom-built 2 MW BDFRG [4]. This control objective has been considered because of the achievable efficiency gain at unity line power factor. Extensive realistic simulation results taking into account the usual practical effects (e.g. transducers' DC offset, noise in measurements, and a PWM power converter model) are presented to support the discussions in this work.

2. Modeling and Operation of the BDFRG

The dynamic model of the BDFRG in arbitrary rotating reference frames, assuming motoring convention, can be represented using standard complex notation as [3]:

$$\left.\begin{aligned} \mathbf{v}_p &= R_p\mathbf{i}_p + \frac{d\boldsymbol{\lambda}_p}{dt} = R_p\mathbf{i}_p + \frac{d\boldsymbol{\lambda}_p}{dt}\big|\theta_{p\,const} + j\omega_p\boldsymbol{\lambda}_p \\ \mathbf{v}_s &= R_s\mathbf{i}_s + \frac{d\boldsymbol{\lambda}_s}{dt} = R_s\mathbf{i}_s + \frac{d\boldsymbol{\lambda}_s}{dt}\big|\theta_{s\,const} + j\omega_s\boldsymbol{\lambda}_s \\ \boldsymbol{\lambda}_p &= \underbrace{L_p i_{pd} + L_{ps} i_{sd}}_{\lambda_{pd}} + j.\underbrace{(L_p i_{pq} - L_{ps} i_{sq})}_{\lambda_{pq}} \\ \boldsymbol{\lambda}_s &= \lambda_{sd} + j.\lambda_{sq} = \sigma L_s\mathbf{i}_s + \underbrace{\frac{L_{ps}}{L_p}\boldsymbol{\lambda}_p^*}_{\lambda_{ps}} \end{aligned}\right\} \quad (1)$$

where the primary and secondary winding are denoted by the subscripts 'p' and 's' respectively, $\sigma = 1 - \frac{L_{ps}^2}{L_pL_s}$ is the leakage factor, and λ_{ps} is the primary flux linking the secondary winding (i.e. the mutual flux linkage).

The fundamental angular velocity and torque relationships for the machine with p_r rotor poles and $\omega_{p,s} = 2\pi f_{p,s}$ applied angular frequencies to the respective 2p-pole primary and 2q-pole secondary windings are [3]:

$$\omega_{rm} = \frac{\omega_p+\omega_s}{p_r} \leftrightarrow n_{rm} = 60.\frac{f_p+f_s}{p_r} \quad (2)$$

$$T_e = \frac{3p_r}{2}(\lambda_{psd} i_{sq} - \lambda_{psq} i_{sd}) \quad (3)$$

$$T_a = J.\frac{d\omega_{rm}}{dt} = T_e - T_L(\omega_{rm}) - F.\omega_{rm} \quad (4)$$

Notice that $\omega_s > 0$ for 'super-synchronous' operation, and $\omega_s < 0$ at sub-synchronous speeds (i.e. an opposite phase sequence of the secondary to the primary winding) in (2) where $\omega_{syn} = \frac{\omega_p}{p_r}$ is the synchronous speed (for $\omega_s = 0$ i.e. a DC secondary) as with a $2p_r$-pole wound rotor synchronous turbo-machine.

It is also worth mentioning that all the ω_p rotating vectors in the primary voltage/flux equations in (1) are in ω_p frame, whilst the corresponding secondary counterparts, including the λ_{ps} components in (3), are stationary in $p_r\omega_{rm} - \omega_p = \omega_s$ frame [3]. Given that λ_p and λ_{ps} in (3) are approximately constant by the primary winding grid connection, torque control can obviously be achieved through the secondary dq currents in the ω_s rotating frame.

Using (2), one can derive the mechanical power equation indicating individual contributions of each BDFRG winding:

$$P_m = T_e.\omega_{rm} = \underbrace{\frac{T_e.\omega_p}{p_r}}_{P_p} + \underbrace{\frac{T_e.\omega_s}{p_r}}_{P_s} = P_p.\left(1 + \frac{\omega_s}{\omega_p}\right) \qquad (5)$$

The machine operating mode is determined by the power flow in the primary winding i.e. to the grid for the generating ($T_e < 0$) regime under consideration, while the secondary winding can either take or deliver real power (P_s) subject to its phase sequence i.e. the ω_s sign; the BDFRG would absorb (produce) $P_s > 0$ at sub- or super-synchronous speeds.

3. Controller System Design

The detailed BDFRG system layout with a generic controller design is presented in Fig. 2. A standard phase-locked-loop (PLL) algorithm, readily available in the *Simulink* library, has been used to retrieve the stationary α-β frame angular positions (θ/θ_p) of the primary voltage/flux vectors from the measured voltages and/or currents. Furthermore, a conventional vector controller with space-vector PWM of the active rectifier has been implemented for control of DC link voltage and unity line power factor [13, 18-20]. The primary real (P) and reactive (Q) power calculations are reference frame invariant and have been done using the stationary frame voltages ($v_{\alpha\beta}$) and currents ($i_{\alpha\beta}$) to avoid unnecessary conversions into their rotating d_s-q_s equivalents, and the use of time-consuming trigonometric functions, allowing so the higher control rates and superior performance in practice. The Q reference is often set to zero ($Q^* = 0$) for the unity primary power factor but can be any other value of interest for a given real power setting (P^*) in power control mode, or the desired angular velocity ω_{rm}^* in variable speed systems. For example, either P^* or ω_{rm}^* may correspond to the Optimum Power Point Tracking (OPPT) of a wind turbine [1], while Q^* was chosen to optimize efficiency of the WECS in this paper.

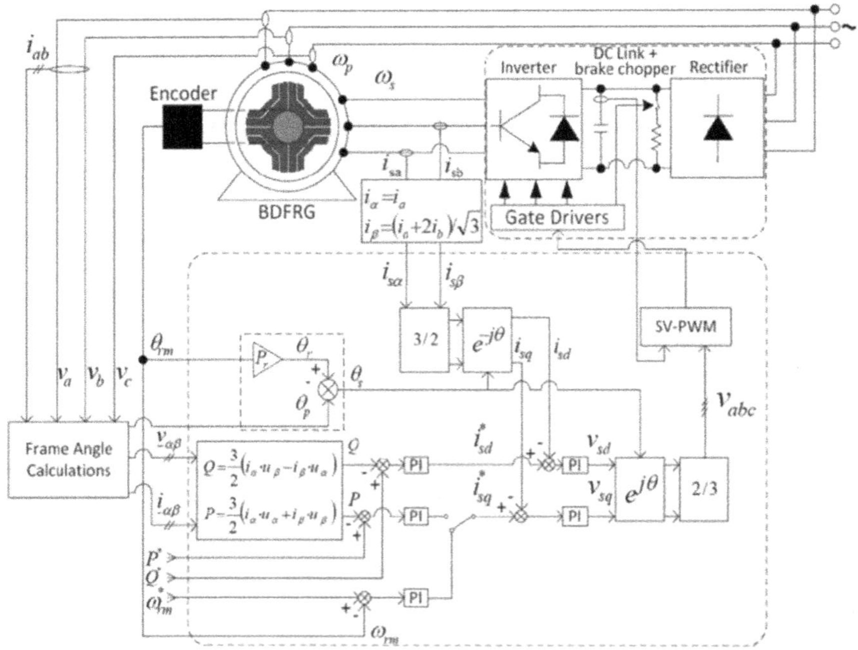

Fig. 2. Structural diagram of the BDFRG drive set-up.

4. Voltage Control (VC) System

The control form expressions can be derived from the BDFRG space-vector model (1) in the natural reference frames, ω_p (i.e. d_p-d_q for the primary winding) and ω_s (i.e. d_s-q_s for secondary winding) rotating frames (Fig. 3), where the respective vector components appear as DC quantities. Substituting for $\mathbf{i}_p$ from the λ_p equation of (1) into $\mathbf{S}_p = \frac{3}{2}\mathbf{v}_p\mathbf{i}_p^*$ would lead to the following relationships for the primary mechanical and reactive power:

$$P_p = \frac{3}{2}\omega_p(\lambda_{psd}i_{sq} - \lambda_{psq}i_{sd}) \tag{6}$$

$$Q_p = \frac{3}{2}\omega_p\left(\frac{\lambda_p^2}{L_p} - \lambda_{psd}i_{sd} - \lambda_{psq}i_{sq}\right) \tag{7}$$

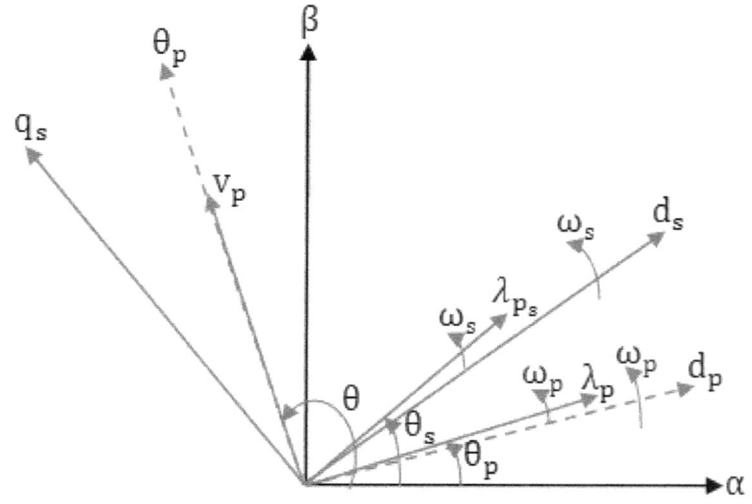

Fig. 3. Identification of primary voltage/flux vectors in a stationary α-β reference frame.

A trade-off of the machine parameter independence is that VC of P_p and Q_p is coupled as both the i_{sd} and i_{sq} secondary current components appear in (6) and (7). The level of this cross-coupling can be reduced by aligning the q_p-axis of the reference frame to the primary voltage vector as shown in Fig. 3. In this case the primary flux vector (λ_p), would be phase-shifted ahead of the corresponding d_p-axis depending on the winding resistance values which are getting smaller with larger machines. Therefore, for the frame alignment choice as in Fig. 3, $\lambda_{p_{sd}} \gg \lambda_{p_{sq}}$ i.e. $\lambda_{p_{sd}} \approx \lambda_{ps}$ so that the approximations of (6) and (7) become:

$$P_p \approx \frac{3}{2}\omega_p\lambda_{ps}i_{sq} = \frac{3}{2}\frac{L_{ps}}{L_p}\omega_p\lambda_p i_{sq} \tag{8}$$

$$Q_p \approx \frac{3}{2}\frac{\omega_p\lambda_p^2}{L_p} - \frac{3}{2}\omega_p\lambda_{ps}i_{sd} \tag{9}$$

$$= \frac{3}{2}\frac{\omega_p\lambda_p}{L_p}\left(\lambda_p - L_{ps}i_{sd}\right) = \frac{3}{2}\omega_p\lambda_p i_{pd} \tag{10}$$

The P_p vs i_{sq} and Q_p vs i_{sd} functions above are nearly linear, which justifies the use of PI controllers in Fig. 2. Therefore, all inductance variations in (9)-(12) can be effectively taken care of by optimal tuning of the PI gains so their prior knowledge is not required for control purposes.

5. PF Operating Principle of the BDFRG Wind Turbine

The preliminary performance of the VC scheme in Fig. 2 has been assessed using the parameters of a large-scale BDFRG [4, 16] in Table 1 for simulation studies. In order to make the simulations as realistic as possible, the following actions have been taken and/or assumptions made: (i) The power electronic models from the *SimPowerSystems* toolbox have been implemented; (ii) High-frequency uncorrelated white noise and unknown slowly varying DC offset have been superimposed to the ideal signals to account for practical effects of the measurement noise and current/voltage transducers errors; (iii) Finally, the rotor position and speed information has been provided by a shaft sensor.

In a typical WECS, the turbine torque on the generator side of the gear-box for the maximum wind energy extraction in the base speed region (i.e. between the minimum 'cut-in', u_{min}, and the rated wind speed, u_r), can be represented as [1, 5]:

$$T_{opt} = \frac{A.\rho.C_p(\lambda_{opt},\gamma).R^3}{2.g^3.\lambda_{opt}^3}.\omega_{rm}^2 = K_{opt}.\omega_{rm}^2 \tag{11}$$

where ρ is the air density, $C_p(\lambda,\gamma)$ is the power coefficient (i.e. the maximum turbine efficiency as $\lambda = \lambda_{opt}$ in this case), $\lambda_{opt} = \frac{R\omega_t}{u}$ is the optimum tip speed ratio for a given wind speed u, ω_t is the turbine rotor angular velocity, γ is the pitch angle (normally fixed to zero to maximize C_p), R the radius of the circular swept area ($A = \pi R^2$), and $g = \frac{\omega_{rm}}{\omega_t}$ is the gear ratio. The shaft torque-speed profile in (4) is of the same form as (11):

$$T_L = -\frac{P_r}{\omega_r}.\left(\frac{n_{rm}}{n_{max}}\right)^2 \approx -19.\left(\frac{n_{rm}}{1000}\right)^2 \text{ kNm} \tag{12}$$

Table1. The BDFRG design specifications.

Symbol	Quantity	Values & Units
J	Rotor inertia	$3.8 kgm^2$
R_p	Primary resistance	0.0375 Ω
R_s	Secondary resistance	0.0575 Ω
L_p	Primary inductance	1.17mH
L_s	Secondary inductance	2.89mH
L_{ps}	Mutual inductance	0.98mH
p_r	Rotor poles	4
P	Primary power	2 MW
n_r	Rated speed	1000 rev/min
$I_{p,s}$	Stator currents	1.5kArms
V_p	Primary voltage	690Vrms
f_p	Supply frequency	50 Hz
Y/Y	Winding connections	-
p/q	Stator poles	6/2

6. Simulation Outcome

The simulation results in Figs. 4-9 have been produced by running the control algorithms in Fig. 2 at 5 kHz switching rate for the IGBT converter. The DC link voltage has been maintained at ≈ 1200 V by the PWM rectifier (i.e. the line-side bridge) supplied at 690 V, 50 Hz. The reference speed trajectory is set as a steep ramp signal suited for dynamically not very demanding wind power applications even under extreme turbulent wind conditions.

The top plots in Fig. 4 show the excellent speed tracking with no overshoot following the start-up period in both super- and sub-synchronous modes of the BDFRG over the limited speed range of 600-900 rpm. The primary electrical power (P) and electro-magnetic torque (T_e) curves reflect (12) for the specific speed settings. Except for a difference in losses, and considering that $\omega_p \approx$ const, P and T_e are directly related as follows from (5) and (6) which explains a close resemblance in their shape. The T_e deviations from the desired profile during the speed transients refer to the acceleration or deceleration torque term in (4) depending on whether the machine is to speed-up ($T_a > 0$) or slow-down ($T_a < 0$). The reasonably accurate and smooth $Q_p \approx 0$ control can be observed to be little affected by the variable speed dependent loading disturbances, which means that the optimum power factor control (OPFC) conditions have been largely met. Note that the outer PI speed control loop is required to ensure effective variable speed operation at optimum tip-speed ratio.

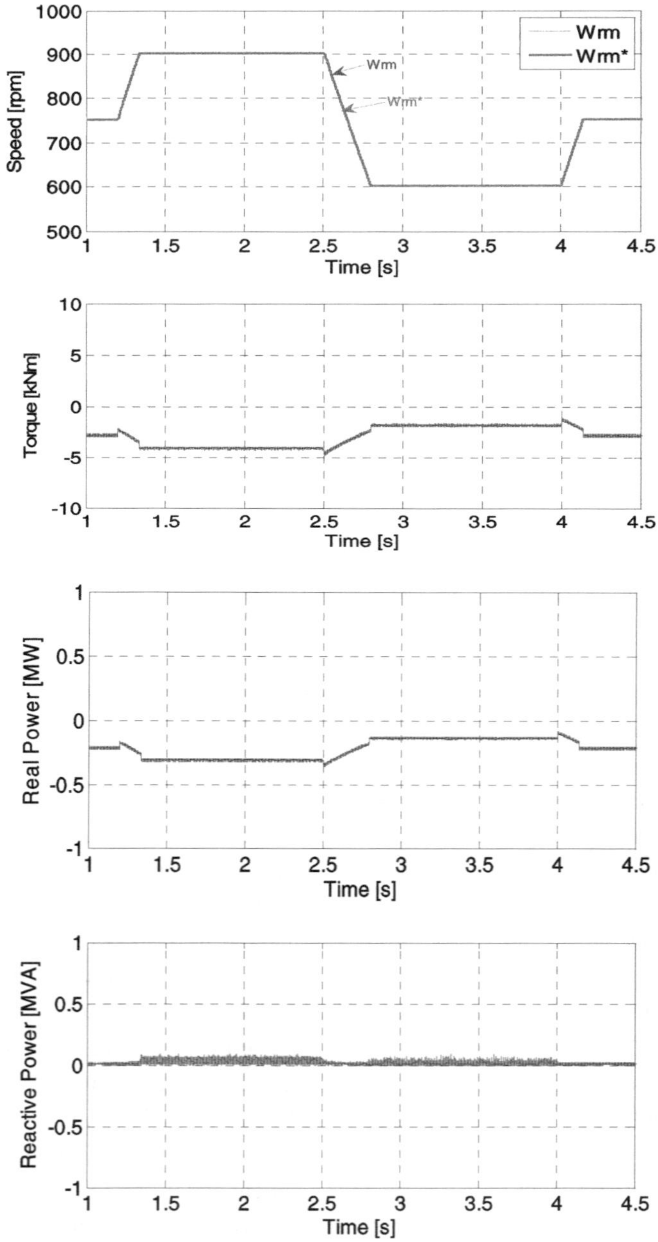

Fig. 4. PF performance of BDFRG in a limited range at synchronous speed (750 rpm).

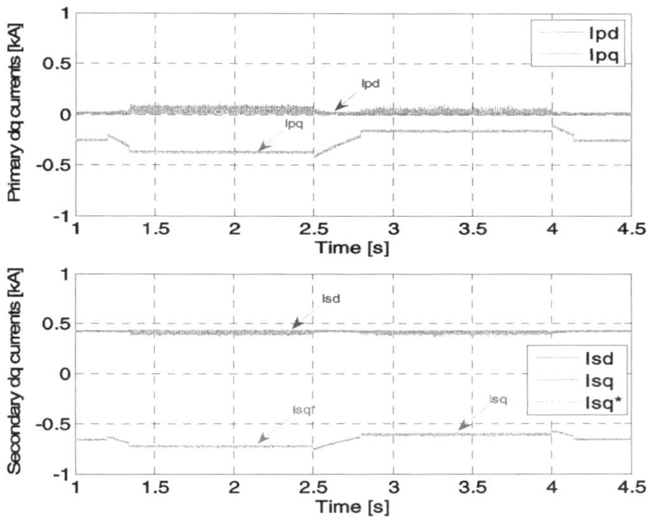

Fig. 5. PF responses of BDFRG current components in the rotating reference frames.

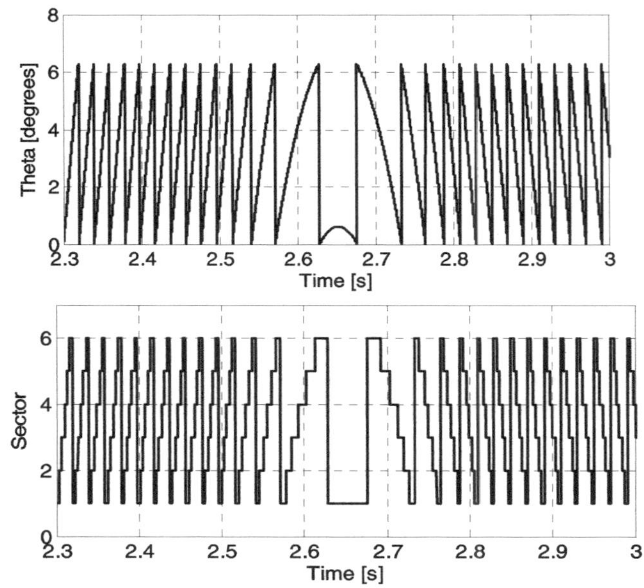

Fig. 6. PWM space Sectors, angular (theta) positions and switching times of active voltage vectors of BDFRG showing phase sequence reversal during the speed mode transition at unity power factor.

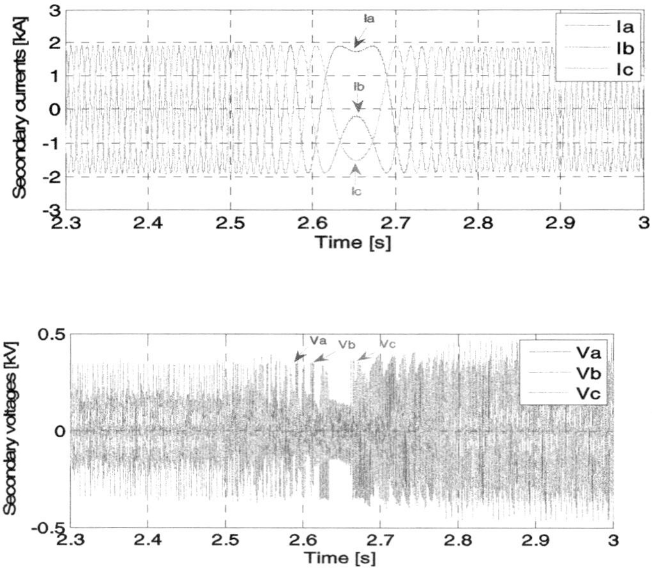

Fig. 7. BDFRG angular (theta) positions and secondary voltage waveforms showing a phase sequence reversal during the speed mode transition at unity power factor.

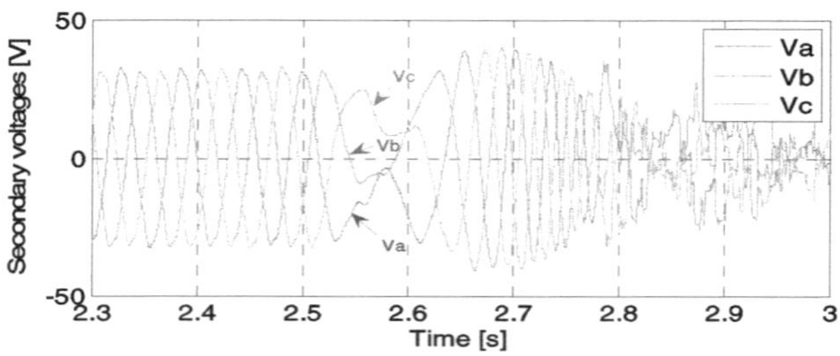

Fig. 8. Enhanced BDFRG secondary voltage waveforms showing a phase sequence reversal during the speed mode transition at unity power factor.

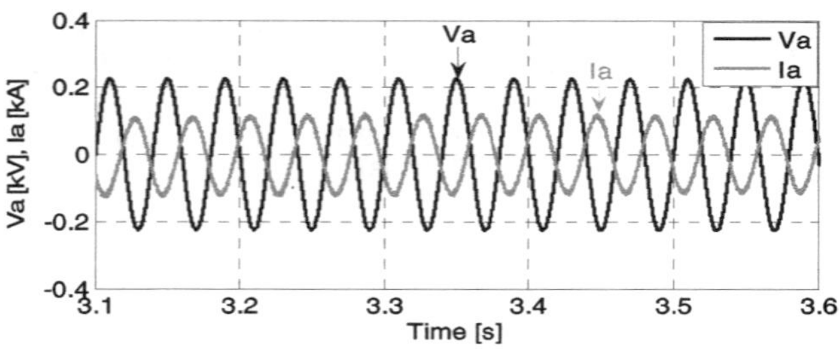

Fig. 9. Primary voltage and current at unity power factor.

The secondary ($i_{sd,q}$) and primary ($i_{pd,q}$) current waveforms in Fig. 5 show no transient over-currents as the PI regulators do not need to be saturated to allow accurate tracking of the desired trajectories for the moderate ramp speed variations. A close link between the active q currents and the real power (torque), as well as the magnetizing d currents and Q, is immediately visible from the respective waveforms. The coupling effects of the i_{sq} clearly manifest themselves as speed (torque) dependent disturbance (e.g. offsets) in the respective non-controllable i_{pd} profiles by analogy to the P and Q scenario in Fig. 4.

Figure 6 illustrates the step-wise PWM angular positions of the modulated secondary voltage vector (v_s), and the respective space sector variations, while Fig. 7 shows the respective secondary currents and voltages, during a speed reduction, from 900 rev/min. to 600 rev/min. That is, the speed mode change, from super- to sub-synchronous. In the super-synchronous mode, v_s rotates anti-clockwise as indicated by the ascending angular steps for the same phase sequence of the windings and $\omega_s > 0$ in (2). The situation is reversed at sub-synchronous speeds when v_s rotates clockwise with the angular steps descending, which comes from the opposite phase sequence of the secondary to the primary winding since $\omega_s < 0$ in (2). Notice that v_s becomes stationary at synchronous speed as the secondary currents are then DC i.e. $\omega_s = 0$ in (2).

Figure 8 shows the enhanced secondary voltage while Fig. 9 shows the respective phase voltage and current waveforms which are π-rad out of phase as expected for unity power factor control (-1 in generating mode) in steady-state period.

7. Conclusions and Perspectives

A novel power factor control (PFC) mechanism for enhanced efficiency of the BDFRG has been carried out and evaluated by comprehensive simulation studies under typical operating conditions. Such vector control implementation can serve as a platform for further research on this emerging and prominent brushless machine topology for applications in large wind turbines and industries, where the cost advantages of partially-rated power electronics and high reliability of brushless structure can be fully utilized.

The enhanced simulation works in Matlab®/Simulink® have clearly demonstrated the immense potential and effectiveness of the controller(s) using the power factor control mechanism. The presented results are more than encouraging and can serve as basis for further research work on this machine.

References

1. M. G. Jovanovic, R. E. Betz and J. Yu, *The use of doubly fed reluctance machines for large pumps and wind turbines*, IEEE Transactions on Industry Applications, vol. 38, pp. 1508–1516, 2002.
2. F. Wang, F. Zhang and L. Xu, *Parameter and performance comparison of doubly-fed brushless machine with cage and reluctance rotors*, IEEE Transactions on Industry Applications, vol. 38, no. 5, pp. 1237–1243, 2002.
3. R. E. Betz and M. G. Jovanovic, *Introduction to the space vector modeling of the brushless doubly-fed reluctance machine*, Electric Power Components and Systems, vol. 31, no. 8, pp. 729–755, 2003.
4. D. G. Dorrell and M. Jovanovic, On the possibility of using a brushless doubly-fed reluctance generator in a 2 MW wind turbine, *IEEE Industry Applications Society Annual Meeting* pp. 1–8, 2008.
5. H. Polinder, F. Van der Pijl, G. de Vilder and P. Tavner, *Comparison of direct-drive and geared generator concepts for wind turbines*, IEEE Transactions on Energy Conversion, vol. 21, no. 3, pp. 725–733, 2006.
6. S. Tohidi, M. Zolghadri, H. Oraee, P. Tavner, E. Abdi and T. Logan, *Performance of the brushless doubly-fed machine under normal and fault conditions*, IET Electric Power Applications, vol. 6, no. 9, pp. 621–627, 2012.
7. F. Barati, R. McMahon, S. Shao, E. Abdi and H. Oraee, *Generalized vector control for brushless doubly fed machines with nested-loop rotor*, IEEE Transactions on Industrial Electronics, vol. 60, no. 6, pp. 2477–2485, 2013.
8. A. Knight, R. Betz and D. Dorrell, *Design and analysis of brushless doubly fed reluctance machines*, IEEE Transactions on Industry Applications, vol. 49, no. 1, pp. 50–58, 2013.
9. T. Long, S. Shao, P. Milliband, E. Abdi and R. McMahon, *Crowbarless fault ride-through of the brushless doubly fed induction generator in a wind turbine under symmetrical voltage dips*, IEEE Transactions on Industrial Electronics, vol. 60, no. 7, pp. 2833–2841, 2013.
10. R. Cardenas, R. Pena, S. Alepuz and G. Asher, *Overview of control systems for the operation of DFIGs in wind energy applications*, IEEE Transactions on Industrial Electronics, vol. 60, no. 7, pp. 2776–2798, 2013.
11. G. Marques and D. Sousa, *Understanding the doubly fed induction generator during voltage dips*, IEEE Transactions on Energy Conversion, vol. 27, no. 2, pp. 421–431, 2012.
12. S. Tohidi, P. Tavner, R. McMahon, H. Oraee, M. Zolghadri, S. Shao and E. Abdi, *Low voltage ride-through of DFIG and brushless DFIG:*

Similarities and differences, Electric Power Systems Research, vol. 110, no. 10, pp. 64–72, 2014.

13. Jude K. Obichere, Milutin Jovanovic and Sul Ademi, *Power Factor Control of Large Doubly-Fed Reluctance Wind Generators*, Lecture Notes in Engineering and Computer Science: Proceedings of The World Congress on Engineering and Computer Science 2015, WCECS 2015, 21–23 October, 2015, San Francisco, USA, pp. 210–215.
14. H. Chaal and M. Jovanovic, *Practical Implementation of sensorless torque and reactive power control of doubly fed machines*, IEEE Transactions on Industrial Electronics, vol. 59, no. 6, pp. 2645–2653, 2012.
15. S. Ademi, M. Jovanovic and J. K. Obichere, *Comparative analysis of control strategies for large doubly-fed reluctance wind generators*, Lecture Notes in Engineering and Computer Science: Proceedings of The World Congress on Engineering and Computer Science 2014, WCECS 2014, 22–24 October, 2014, San Francisco, USA, pp. 245–250.
16. S. Ademi and M. Jovanovic, *Control of emerging brushless doubly fed reluctance wind turbine generators*, in large scale renewable power generation, series on Green Energy and Technology, eds. J. Hossain and A. Mahmud (Springer, Singapore, 2014), pp. 395–411.
17. Milutin Jovanovic, Sul Ademi and Jude K. Obichere, *Comparisons of Vector Control Algorithms for Doubly-Fed Reluctance Wind Generators*, Transactions on Engineering Technologies, World Congress on Engineering and Computer Science 2014, eds. Haeng Kon Kim, Mahyar A. Amouzegar and Sio-long Ao (Springer, Dordrecht, 2015) pp. 85–99.
18. Jude K. Obichere, Milutin Jovanovic and Sul Ademi, *Power Factor Control of Large Doubly-Fed Reluctance Wind Generators*, Lecture Notes in Engineering and Computer Science 2015, WCECS 2015, 21–23 October, 2015, San Francisco, USA, pp. 210–215.
19. Jude K. Obichere, Milutin Jovanovic and Sul Ademi, *Improved Power Factor Controller for Wind Generator and Applications*, Engineering Letters, vol. 24, no. 2, pp. 125–131, 2016.
20. Jude K. Obichere and Milutin Jovanovic, *Optimum Power Factor Control of Large Doubly-Fed Reluctance Wind Generator Applications*, Lecture Notes in Engineering and Computer Science: Proceedings of The World Congress on Engineering 2016, WCE 2016, 29 June–1 July, 2016, London, U. K., pp. 291–296.
21. S. Ademi and M. Jovanovic, *Vector control methods for brushless doubly fed reluctance machines*, IEEE Transactions on Industrial Electronics, vol. 62, no. 1, pp. 96–104, 2015.

WIRELESS STRUCTURAL HEALTH MONITORING SYSTEM BY CAPACITOR SENSING TECHNIQUE

A. H. M. ZAHIRUL ALAM*, NURUL ARFAH CHE MUSTAPHA†, NOOR SYAMIRA ASYIQIN AMIR HASSAN‡, MD RAFIQUL ISLAM§, SHEROZ KHAN¶, and AMELIA WONG AZMAN‖

Department of Electrical and Computer Engineering, Int. Islamic University Malaysia, P.O. Box 10, 50728 Kuala Lumpur, Malaysia
**zahirulalam@iium.edu.my*
†n.c.2013@ieee.org
‡syamira.amir@gmail.com
§rafiq@iium.edu.my
¶sheroz@iium.edu.my
‖amy@iium.edu.my
http://staff.iium.edu.my/zahirulalam/v3/

MUHAMMAD ABU EUSUF

Department of Civil Engineering, International Islamic University Malaysia, P.O. Box 10, 50728 Kuala Lumpur, Malaysia
abueusuf@iium.edu.my

Structural health monitoring (SHM) is a method of localizing damage in infrastructures regarding structural strain, vibration or movements due to external contributing factors which mainly occur as a result of natural disasters. The need to analyze such occurrences can prevent potential damage of massive scale. A sensing system with high sensitivity is needed for reliable structural health monitoring. A capacitive sensing circuit is designed using off-the-self-components and transmit data wirelessly using XBee. The circuit can detect small changes of capacitance variation due to structure vibration.

Keywords: Capacitor Sensor; Differential CVC; Instrument Amplifier; Structural Health Monitoring System; Wireless Monitoring.

1. Introduction

Civil structures involve the meticulous construction of buildings, roads, bridges and tunnels regarding design and maintenance to prolong the lifetime of such structures. Often, these structures as well as its integrity also define a country's socioeconomic well-being. The larger the structure, the harder it is to monitor its structural health and the more expensive it is to keep it under constant maintenance. Natural disasters such as earthquake, typhoon and landslides are major contributing factors to the decline in structural life expectancy. This increases the demand for a thorough implementation of the structural health monitoring system as tragic incidents such as collapsed bridges and cracked structures have a direct relationship to public safety. If neglected, it can lead to huge amounts of casualties apart from causing a country's economy to suffer.

The ultimate aspect of damage analysis and localization in infrastructures can be made by diagnosing the structural deviation from its original and optimal performance [1]. From henceforth, the analysis as to how much deviation has occurred and what type of action that needs to be taken can immediately follow with great regards to the safety of the structure. Over time, researchers have conducted the study of suitable materials that can be used to manufacture accurate sensing abilities. The SHM systems as it usually involve the monitoring of massive infrastructures almost simultaneously. Apart from that, it is also important to comprehend and interpret the type of data that is captured by the SHM systems. Vibration and inclination measurements are examples of data that can be captured by sensors and are vital in SHM systems.

Sensors used in SHM systems should be appropriately and strategically placed over a structural system while data acquisition is needed to monitor receiving signals. Sensors that are high in sensitivity allows the collection of more accurate data. This means that even the subtlest changes can be recorded and monitored. Immediate action can be taken in response to any detection that might lead to negative outcomes. Needless to say, the method to achieve reliable data collection is essential for an SHM system to function efficiently.

Moreover, a wide range of sensors are used to analyze key factors affecting structural integrity. These sensors monitor parameters such as movement and strain of structures. Currently, there are different types of sensors such as electrical resistance, vibrating wire sensors and distributed sensors. These sensors help to detect even the smallest movements that are almost impossible to detect with the naked eye. Additionally, these sensors are used within methods that range from traditional strain monitoring techniques such as fiber optics, acoustics and ultrasonic to more sophisticated techniques such as 3D laser scanning and digital imaging techniques.

In recent years, there has been an increase in interest in the use of wireless networks for many SHM systems due to the presently ever-evolving technological era [2–4]. The applications of such wireless networks are associated with several challenges that need to be taken into consideration upon implication. Such issues that need to be taken into consideration concern the cost of maintenance and the sensing reliability of the system. This is due to the fact that the implementation of a guided system includes high costs of installation and is an industrially arduous work, including the need for frequent maintenance and inspection. More than that, if a guided system is used instead of a wireless network, this cost will increase significantly over massive structures. Hence, the demand for a wireless network gives the industry an opportunity to decrease costs in many aspects and is a distinguishable advantage to many.

Previous works have discussed a variety of sensors in an assortment of network configurations. However, very few research was done on a capacitor sensor system using wireless networks. Research in [5] have implemented capacitor sensor system in the health monitoring application, and others have designed suitable for high frequency range [6–10]. Thus, this project focuses on the construction of a wireless capacitor sensor system that is to be implemented for structural health monitoring. An extended work done in [11] is presented in this paper.

2. The Design

The block diagram of the wireless structural health monitoring system is shown in Figure 1. It consists of a capacitance sensor for detecting any changes in structure which can be measured by the changes in the capacitance. The changes of capacitance are reflected in terms of voltage. The captured voltage is monitored through Arduino microcontroller. All transmitting and receiving of data is via Xbee.

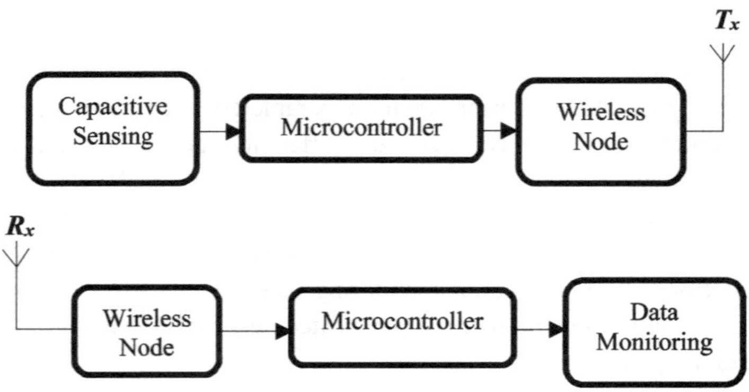

Fig. 1. Block diagram of the structural health monitoring system.

The circuit diagram of the capacitor sensor network is shown in Figure 2. The circuit consists of a reference capacitor, C_1 and a sensing capacitor, C_3. The reference capacitance, C_1 is chosen to be equal to the sensing capacitance, C_3 value of 170 pF.

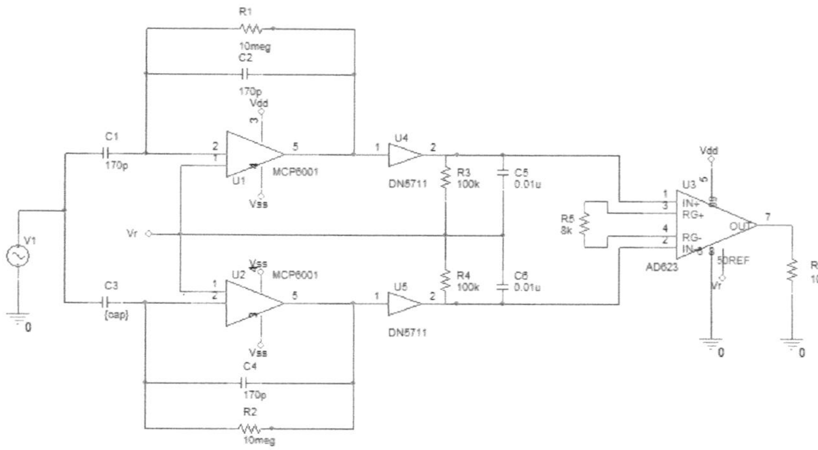

Fig. 2. Capacitive sensing circuit.

The first part of circuit in Figure 2 act as a differential sensing amplifier using operational amplifiers followed by Schottky rectifier with RC filter. The non-inverting terminals of both the Op-Amps are shorted and connected to a reference voltage, V_r which act as a reference voltage of ADC of Arduino input. The design starts with deriving the transfer function at one of the output sensing amplifiers, V_{sense} to V_1. The transfer function of the sensing amplifier is

$$V_{sense} = \left[1 + sC_3 \frac{R_f}{1 + sR_f C_f}\right] V_r - \left[sC_3 \frac{R_f}{1 + sR_f C_f}\right] V_1 \cdot \quad (1)$$

Where feedback resistor and feedback capacitor are $R_f = R_1 = R_2$ and $C_f = C_2 = C_4$ respectively. Several conditions may exist when the reference voltage, V_r is supplied to the non-inverting input of the amplifier. Based on the Figure 2, the following cases may exist:

(1) Case I: $V_r = 0$

$$\left|\frac{V_{sense}}{V_1}\right| = -sC_3 \frac{R_f}{1 + sR_f C_f} \cdot \quad (2)$$

(2) Case II: $V_r = V_1$

$$\left|\frac{V_{sense}}{V_1}\right| = 1 \cdot \quad (3)$$

(3) Case III: $V_r < V_1$ when $V_r = (1 - m)V_1$ where $(0 < m < 1)$

$$\left|\frac{V_{sense}}{V_1}\right| = 1 - \left(1 + sC_3 \frac{R_f}{1 + sR_f C_f}\right) m \cdot \tag{4}$$

(4) Case IV: $V_r > V_1$ when $V_r = nV_1$ where $(n > 1)$

$$\left|\frac{V_{sense}}{V_1}\right| = n\left(1 + sC_3 \frac{R_f}{1 + sR_f C_f}\right) - sC_3 \frac{R_f}{1 + sR_f C_f}, \tag{5}$$

where m and n is the multiplier number of the excitation input voltage, V_1.

This work has used input voltage, $V_1 = V_{amp} \cos(\omega t + \varphi)$ where V_{amp} is the amplitude of the input voltage which is equal to 400 mV, applied to the sensing amplifiers. Input frequency is 40 kHz. Reference voltage, V_r is set at half of the supply voltage, $V_{dd} = 3$ V or in this work, $V_r = 3.75V_1$. For (Case IV: $V_r > V_{amp}$), the output of the diode U5 is equal to Eq. (6)

$$V_{Output\_U5} = 2.75\,\hat{V}_1 \cdot \frac{(C_3 + \Delta C)}{C_f} \cos \omega t - V_{diode\_U5} \tag{6}$$

The output of the diode U4 is

$$V_{Output\_U4} = \left[2.75\;\hat{V}_1 \cdot \frac{C_1}{C_f}\right] - V_{diode\_U4} \tag{7}$$

Where the $\hat{V}_1$ is the amplitude voltage after demodulation. Notice that in both Eq. (6) and (7), of U4 and U5 diode voltages. Since similar types of diodes were used in this work, it can be expected that the different of both voltage drop diode U4 and U5 is very small. This two voltage drops can be cancelled out. The outputs of the two diode rectifiers are then fed into an instrumentation amplifier which senses the differential change of these two outputs as expressed in Eq. (8) [6].

$$V_{diff} = \pm \frac{C_3 + \Delta C}{C_f} \cdot \alpha \cdot V_{amp} \cdot \tag{8}$$

Where α is the attenuation of the diode rectifier and V_{amp} is the amplitude of the input voltage, $V_1 = V_{amp} \cos(\omega t + \varphi)$ applied to the sensing amplifiers. For (Case IV: $V_r > V_{amp}$), the differential change can also be expressed as

$$V_{diff} = 2.75\ \hat{V}_1 \left[\frac{(C_3 + \Delta C)}{C_f} \cos \omega t - \frac{C_1}{C_f} \right] \tag{9}$$

The output of the differential amplifier is given by Eq. (10)

$$V_{out} = G \cdot V_{diff}. \tag{10}$$

Where G is the differential amplifier gain which can be controlled by the resistance R_5. The relevance of Eq. (10) is when there is no capacitance change in C_3, there is no movement at the sensor, $\Delta C = 0$. Hence, the frequency input is 0. When C_3, C_1 and C_f are all set at nominal value of 170 pF, the result of output voltage, V_{out} is equal to 0 V. The value of G can be selected so that the output voltage is not exceeding the supply voltage in order to avoid saturation.

The following components MCP6004 Op-Amps, DN5711 Schottky diodes and AD 623 instrumentations are used for the circuit shown in Figure 2. The sensing capacitance is designed using two parallel copper plate of dimension 10 cm × 10 cm with a separated gap. The measured value of capacitance is 170 pF. Arduino UNO is used as an ADC to measure the instrument amplifier output voltage and the Arduino is attached to XBee module to transmit the data. Another Xbee is connected to the personal computer to receive data as indicated in the block diagram of Figure 1. The complete implementation of the circuit is shown in Figure 3.

Fig. 3. Complete implementation of the sensing circuit.

3. Result and Discussion

The reference capacitance is varied from 170 pF to 200 pF with the vibration. As vibration occurs in the structural body causes to vary the distance between parallel plate capacitance. Simulation has been done using PSpice simulator to observe the behavior of the output of the instrumentation amplifier. The output voltage of the instrument amplifier with the variation of the sensing capacitance is shown in Figure 4.

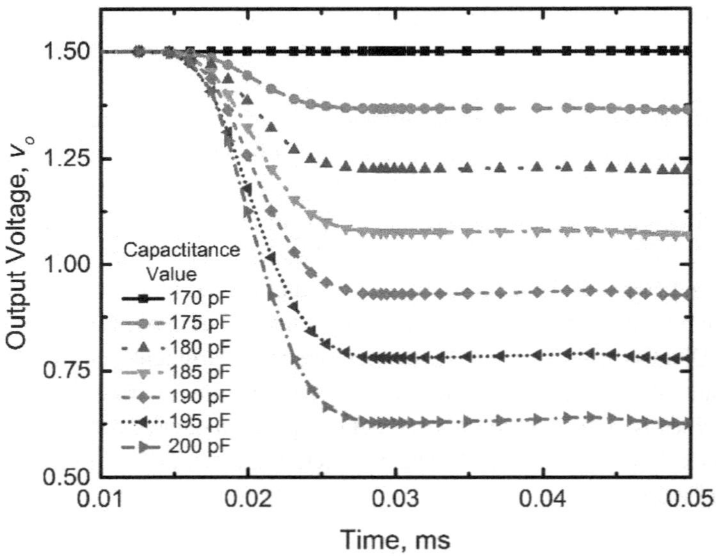

Fig. 4. Instrument amplifier output voltage with the variation of the sensing capacitor.

It is observed that the output voltage is 1.5 V, which is equal to the reference voltage when the sensing capacitor value is 170 pF which is equal to the reference capacitance. The output voltage decreases as the value of the sensing capacitance increases. The output voltage of the instrumentation amplifier can be increased by interchanging the reference capacitor, C_1 and the sensing capacitor C_3. The effect of the interchanging capacitance is shown in Figure 5.

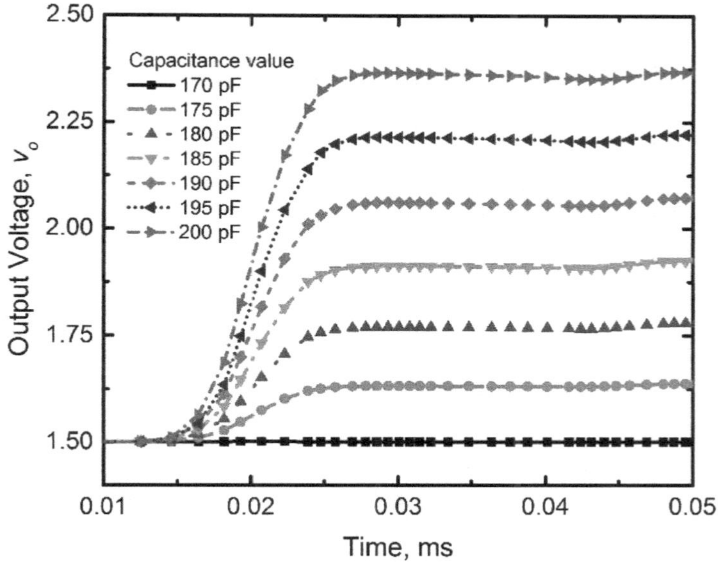

Fig. 5. Instrument amplifier output voltage with the variation of the sensing capacitor by interchanging sensing and reference capacitor.

Figure 6 shows the variation of the output voltage of the instrument amplifier with the change of the sensing capacitance on the reference capacitance. It shows that the output voltage increases or decreases linearly with the change of the capacitance depending on the position of the sensing capacitor as predicted by Eq. (6). It is observed that the output voltage change at the rate of 25 mV with the change of 1 pf capacitance for the differential amplifier gain, G of 13.5.

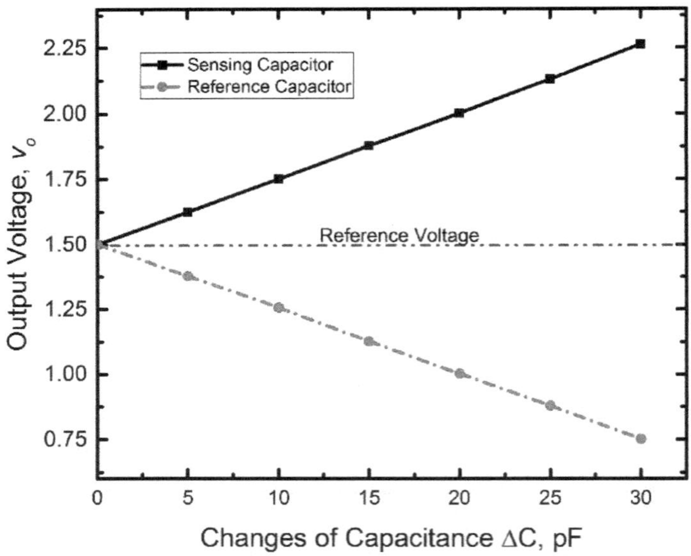

Fig. 6. Instrument amplifier output voltage with the changes of capacitances.

The output of the instrument amplifier is fed to the ADC pin of Arduino with Xbee. Arduino is properly configured to read the exact voltage of the amplifier and transmit data by Xbee. The other receiver Xbee also configured properly to receive data. It is found that the Xbee received proper data corresponding to the sending data.

4. Conclusion

A simple capacitive based structural health monitoring system has been proposed. The circuit can measure a very small capacitance change. The proposed circuit can give an output of 25 mv for every pf change. The measured data can easily transmit using Arduino microcontroller based system. Xbee is successfully used for wireless data transmission. Many sensor nodes are required for monitor a large structure.

Acknowledgments

This work was supported by the Ministry of Higher Education Malaysia under Exploratory Research Grant Scheme (ERGS), ERGS13-021-0054.

References

1. H. M. Milenski, "Technical Assistance for Remote Sensor Placement Related Projects for Structural Health Monitoring," *Report Los Alamos National Laboratory*, USA (2014) doi: 10.2172/1159044
2. C. Liu, J. Teng and N. Wu, "A Wireless Strain Sensor Network for Structural Health Monitoring," *Shock and Vibration*, 2015, http://dx.doi.org/10.1155/2015/740471
3. M. Bocca, J. Toivola, L. M. Eriksson, J. Hollmén and H. Koivol, "Structural Health Monitoring in Wireless Sensor Networks by the Embedded Goertzel Algorithm," *IEEE/ACM Second International Conference on Cyber-Physical Systems*, 12-14 April 2011, pp. 206–214.
4. D. Zhou, D. S. Ha and D. J. Inman, "Ultra Low-Power Active Wireless Sensor for Structural Health Monitoring," *Smart Structures and Systems*, Vol. 6, No. 5-6, 2010, pp. 675–687.
5. N. A. C. Mustapha, A. H. M. Z. Alam, S. Khan and A. W. Azman, "Efficient Capacitance Sensing for Wireless Health Monitoring Sensing," *IIUM Engineering Journal, 2016*, Vol. 17, No. 2, [in press].
6. M. R. Haider, M. R. Mahfouz, S. K. Islam, S. A. Eliza, W. Qu and E. Pritchard, "A Low-Power Capacitance Measurement Circuit with High Resolution and High Degree of Linearity", *Proc. of the 51st Midwest Symposium on Circuits and Systems, 2008. MWSCAS 2008.* 10–13 Aug. 2008, pp. 261–264.
7. A. H. M. Z. Alam, N. Arfah, S. Khan and M. R. Islam, "Design of Capacitance to Voltage Converter for Capacitive Sensor Transducer," *American Journal of Applied Sciences*, Vol. 7, No. 10, 2010, pp. 1353–1357.
8. N. Arfah, A. H. M. Z. Alam and S. Khan, "Capacitance-to-Voltage Converter for Capacitance Measuring System", *2011 4th Int. Conf. on Mechatronics (ICOM)*, 2011, pp. 1–4.
9. N. Arfah, A. H. M. Z. Alam and S. Khan, "CMOS Op amp Testing for Capacitive Measuring Systems Application," *2011 IEEE Regional Symp. On Micro and Nano Electronics*, 2011, pp. 177–181.
10. N. A. C. Mustapha, A. H. M. Z. Alam, S. Khan and A. W. Azman, "Capacitance Sensing and Energy Harvesting for Wireless Health Monitoring System," in *2016 International Conference on Intelligent Systems Engineering (ICISE)*, 15 Jan–17 Jan 2016, pp. 326–329.

11. A. H. M. Zahirul Alam, N. S. A. A. Hassan, N. A. C. Mustapha, M. Rafiqul Islam, K. Sheroz and M. A. Eusuf, "Wireless Capacitor Sensing for Structural Health Monitoring," *Lecture notes in Engineering and Computer Sciences: Proceedings of The World Congress on Engineering 2016, WCE 2016*, 29 June–1 July, 2016, London, UK, pp. 319–321.

AUTOMATED VERMICAST PRODUCTION WITH ANDROID APPLICATION

MARY ANNE A. MACABUHAY
Electronics Engineering Department, Don Honorio Ventura Technological State University,
Bacolor, Pampanga, Philippines
ma_macabuhay@yahoo.com

MA. TERESA R. ABELLERA
Electronics Engineering Department, Don Honorio Ventura Technological State University,
Bacolor, Pampanga, Philippines
tere.abellera@gmail.com

ADEL E. TICSAY
Arduino Group
Bacolor, Pampanga, Philippines
adelticsay@yahoo.com

Vermicomposting is the process of breaking down biodegradable matter by earthworms to convert the contained nutrients in the organic matter to vermicast. In this paper, the proponents introduced the development of an automated production of vermicast. Four subsystems were designed namely: irrigation, sensor network, worm migration, and communication subsystem to minimize human intervention. The communication system involved the use of 2.4 Ghz. The study uses Arduino uno and Arduino mega microcontrollers, an android phone and liquid crystal display (LCD) for monitoring. The automated project improved the manual process of vermicomposting by eliminating 4 processes from the manual system.

Keywords: Arduino; Soil Moisture; Vermicast; Vermicomposting.

1. Introduction

Organic farming is a classification of agricultural production that aims to produce food and fiber. It was conceived to be the best alternative production method for a healthier life, since it prohibits the use of synthetic pesticides or fertilizers. This is expected to have high rise in demand. For organic farming, castings of earthworms also known as vermicomposting is an excellent soil enhancer and bioactive high quality fertilizer. Vermicomposting is the process of breaking down biodegradable matter by earthworms to convert the contained nutrients in the organic matter to vermicast.

Eudrilus eugeniae or commonly known as African night crawler is an earthworm species indigenous to Africa but extensively bred in the USA,

Canada, Europe and Asia [1]. It has high-production rates and is capable of decomposing large quantities of organic wastes quickly and incorporating then to topsoil [2]. It shows preferences for high temperature, with maximum biomass production occurring at 25°C–30°C, while growth rates were very low at 15°C [3]. It can tolerate moisture between 70%-80%, the optimum being 80%-82%. It can also tolerate pH levels from 5-9, but they move on more acid material, with pH material of 5 [4].

There are around 4000 species of Earthworm worldwide and 400 of them can be found in the Philippines. In 1982 Dr. Otto Graff introduced African Nightcrawler (ANC) in the Philippines which originated in West Africa. In tropical country like the Philippines, ANC increases rapidly due to favorable weather condition. The vermicast production is an emerging farming observed in the developing country like the Philippines and is still in manual system. Region 3 and Region 12 of the country are agriculture-based industry area, where vermiculture and vermicast production industry displays good potentials in the local market. Vermiculture is the culture of earthworms. The goal is to continually increase the number of worms in order to obtain a sustainable harvest [5].

The authors conducted site survey of a local vermicast farm industry in Region 3 of the country (Fig. 1). The manual process starts by laying of blocks as perimeters for compost. This also serves as the structure for the worm bed. Substrates are then prepared and placed in the whole area of the worm bed. Upon equal distribution of substrates, a net is then placed before placing the earthworms. The net serves as the means in separation of the worm and the harvest vermicast. The vermicompost is watered three times a day and once the compost becomes a pellet-like mixture, the worms are manually removed from the vermicompost and vernicast harvesting begins. Turning organic wastes into casts takes 22–32 days, depending on density of waste and earthworm maturity-45 days [6]. African Night Crawler can consume 1 kilogram of organic material, while research suggests that the earthworms ingest about 75% of their body weight/day and a 0.2 gram worm eats about 0.15 gram per day.

Fig. 1. Manual system vermicast production. Blocks are used as fenced for the worm bed and substrates.

Regular composting requires 30 to 40 days, and additional 3 to 4 months of curing. However, the advantage of vermicast is that it does not require to be cured before it is used as fertilizer [7].

An automated system of producing organic fertilizer can greatly increase the development of the current agricultural technique. One example is found at a cabbage farm at Murcia, Sapin, where wireless sensor network is used in detecting different agricultural factors such as water content, temperature and salinity of soil [8]. The production of vermicompost for organic farming still demands further study and improvement in the developing country like. From the two site surveys on vermicast farm that the authors conducted, manual system of vermicomposting are being practiced. Hence, it is evident that there is limited application of electronics technology adapted with the existing process. The study aims to provide an automated system for vermicast production that may improve the whole concept of organic farming.

2. Concept and Subsystem

The design and development of the study consists of considering the control variable (Fig. 2). ANC was considered as major concern for the vermicast production. In which, their environmental conditions to sustain their growth rates is vital. The authors consider soil moisture as primary considerations followed by the soil temperature, light intensity and pH levels.

Two Arduino microcontrollers were used in the project. Arduino 1 was used to receive signals from the input sensor networks and processed these signals to trigger the desired output. Arduino 2 was used as a timer for the gate of the worm bins. A simplex communication between the two Arduino was established to coordinate the time programmed to open and to closed the worm bin's gate.

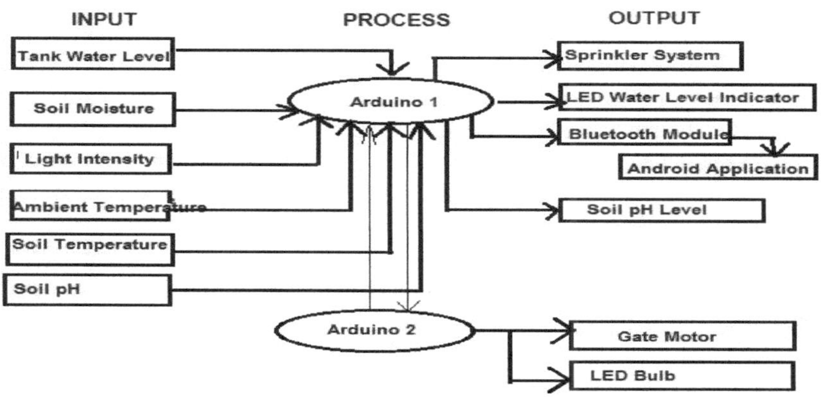

Fig. 2. Conceptual Framework of the study. Two Arduino microcontroller were used for the study to control two worm bins and input parameters.

2.1. *Soil Moisture and Soil Temperature*

Soil moisture and soil temperature sensors monitor the status of the vermicompost in the worm bin. If the soil temperature reached 33 degrees Celsius and soil moisture goes below 60%, the sensors will send the data to the arduino microcontroller for condition checking. The set condition will then trigger the water pump to activate the sprinkler to maintain the soil temperature and soil moisture, thereby avoiding the death of the earthworms [7].

Moisture content of 40% and 50% slows decomposition [9]. Moisture level must be maintained at around 50% so that microbial activity is high and the food matter is easy to feed up [10]. However, in Philippine settings, worms thrive best with moisture content with 60-80 %. The more worms stocked, the faster the production of casting when substrate is fully composted in 4 to 5 weeks (25-35 days) the vermicompost can be harvest. The data that the sensors detect is also processed by the Arduino mega and is transmitted via Bluetooth to android phone for monitoring purposes.

2.2. *Sensor Network*

The study uses five kinds of sensors namely: soil moisture sensor, ambient and soil temperature sensor, light intensity sensor and water level sensor.

The soil moisture sensor that was used consists of two electrodes. Prior to embedding it into the vermicompost, these electrodes are coated with plaster of Paris to enable it to absorb water from the vermicompost before exposing the sensors to the soil moisture. This eliminates fluctuations of value read by soil moisture sensor. The sensor was programmed to read values from 0%-100% soil moisture.

Determination of soil moisture minimum and maximum threshold was done using two pint calibration. A ratio of 1:1 was used, that is, 3500 ml of soil was sun-dried for 48 hours. A measured value of 0% soil moisture indicates the baseline or minimum soil moisture. An additional of 3500 ml of water mixed with soil indicates maximum soil moisture or 100% soil moisture content. The sensor was programmed from the range of 0%-100%.

The soil and ambient temperature sensors are both thermocouple nodes that are connected to the Arduino module. Thermocouples are comprised of two dissimilar conductors joined at one end to form a hot junction which produces a voltage. As the junction is heated or cooled the voltage changes and is converted to return the temperature.

The light-dependent resistor (LDR) gives out an analog voltage when connected to supply voltage of Vcc= 5V, which varies in magnitude in direct proportion to the input light intensity on it. This output analog voltage from LDR is then fed to the analog input pin of the arduino. The built-in Analog to Digital Converter converts the analog voltage from 0V-5V into a digital value in the range of 0-1023. Thus, when there is sufficient light in its environment or on its surface, the converted digital values read from the LDR through the arduino will be in the range of 800-1023.

The authors used ultrasonic sensor for water level, which measures the height of the water inside the tank.

2.3. *Water Irrigation System*

For the control of the sprinkler system, solenoid water valves were used (Figs. 3 and 4). Instead of creating pressure by use of a pump, the pressure generated from the elevation of the water tank is utilized instead, and water flow is enabled or disabled by the solenoid water valves. The water valves operate with 12 Volts, 1 ampere and can withstand 0.02 to 0.8 Mega Pascal of water pressure. The water tank is a plastic drum having a diameter of 60.20 centimeters and a height of 82.55 centimeters. The total volume of water it can hold is 234.95 liters. The sprinkler system is responsible for watering the compost when triggered by the system to maintain the preferred soil moisture and soil temperature for the compost.

Fig. 3. Water sprinkler system. The design ensures well distributed water over the area of the vermicompost.

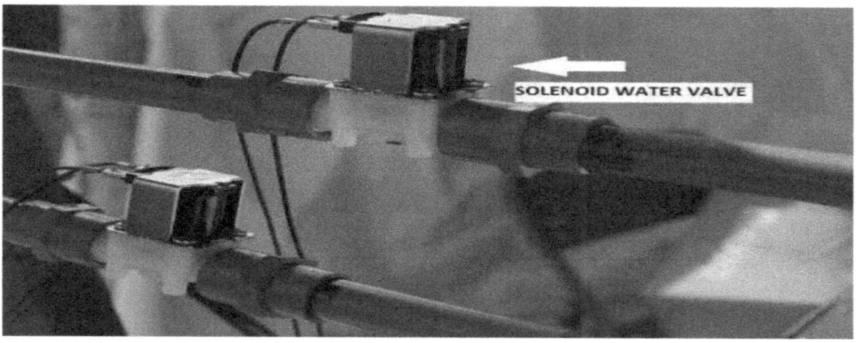

Fig. 4. Solenoid water valve. Compatible with the used of arduino microcontroller, cheaper than submersible water pump and synergizes with water pump.

2.4. *Worm Migration System*

Two worm bins were prepared for faster worm migration (Figs. 5 and 6). The first bin was prepared for the first vermicomposting. Each bin has a volume of 0.4 m^3 with an ideal harvest of 0.083 m^3 to 0.1 m^3 vermicast. The second worm bin was made ready for worm migration for the next vermicomposing. Earthworms vacate the place if they find no food to consume. In this case, when they have consumed the decomposed materials from the first bin, worms will get attracted to the next worm bin because of the presence of the decomposed. It was observed that worms took 20 days to consume the decomposed. The Arduino 2 will trigger the gate to open after 20 days, and to assure the transferring of worms between the bins, a lamp will light-up during worm migration. ANC does not like lights, for this reason, the lamp helps in worm migration. The lamp would turned off once the gate is closed. The evaluated worm bin 1, consists of produced vermicast which is ready for harvest.

Fig. 5. Automated vermicast system. The study has 4 subsystems: water irrigation, migration, sensor network, and communication subsystem.

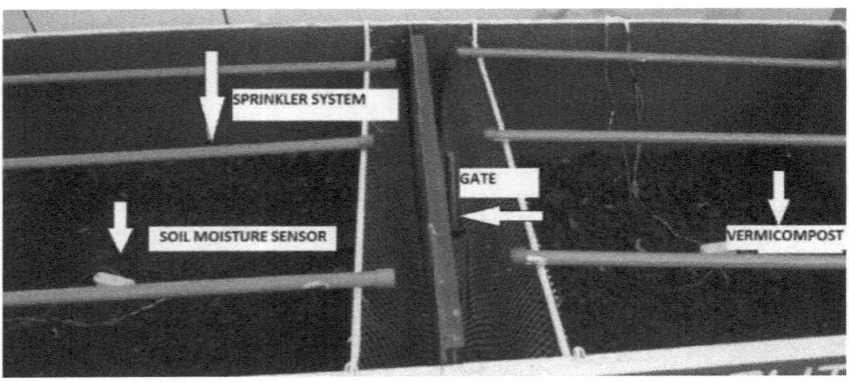

Fig. 6. Worm bins with gate controlled. Two worm bins were designed for worm migration to eliminate the process of manual separation of worms with the produced vermicast.

The required motor used for the project needs a forward and reverse mode of rotation. The motor used for this project has a speed of 2-3 revolution per minute, 12V, 6A and brush commutation. The motor was designed for high torque and low speed applications.

2.5. *Bluetooth and Android*

The Bluetooth connection is provided by the Bluetooth module connected to the arduino 1 (Fig. 7). It is a BC417, single chip radio and baseband IC for Bluetooth 2.4 GHz systems including enhanced data rates (EDR) to 3 Mbps. Bluetooth V2.0 is used by the module, using a short-wavelength ultrahigh frequency (UHF) band from 2.4 to 2.485 GHz, and has a physical range of 10 to 100 meters. This version of Bluetooth is used in the project in conjunction with the android app so that the user will have a monitoring application displaying the parameters the project is processing besides the Liquid Crystal D display when the user is near the project itself.

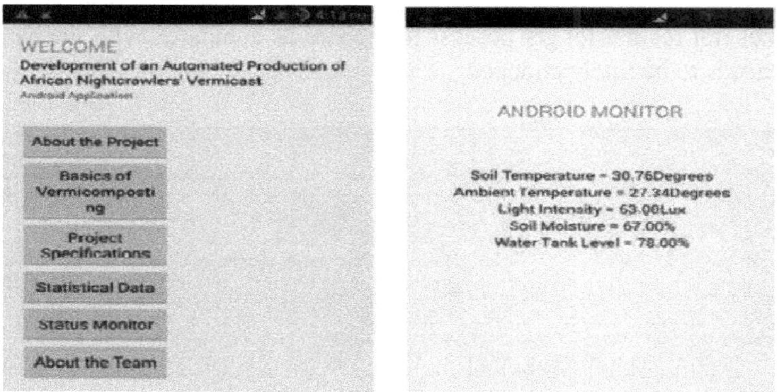

Fig. 7. Developed android application (left) & Android Monitoring(right). The Arduino 1 sends data to be viewed using Liquid Display (LCD) and android application

Data received by arduino1 is sent via bluetooth to the android phone. This was done by using the android application APK file to view the data from Arduino through android phone.

3. Test and Analysis of Results

The development of the system includes several test performance to undertake. Experiments and test measurements were performed on hardware and selection of best pre-decomposed substrate. These tests were performed based on the following hierarchy: Individual test prior to system integration, Subsystem test and system integration test, respectively.

3.1. *Selection of Best Pre-Decomposed Substrate*

Prior to performance of each tests, selection of the best pre-decomposed must first be considered (Fig. 8). These two pre-decomposed are chopped banana stalk and sugar bagasse. Test results on soil moisture and soil temperature of these materials as pre-decomposed substrate were measured and data were taken every hour for 3 days. Statistical treatment used was two-tailed t-test to determine if there exists difference between materials. The critical region was found at t(46, 0.05) = 2.013 with t Stat value of –.702. Therefore, the mean soil moisture of materials has no difference.

Test on soil temperature of both materials shows that pre-decomposed substrate with sugar bagasse has a mean soil temperature of 25.68°C while substrate with chopped banana stalk has 25.08°C. There is a closeness of soil temperature value for both materials. Statistical treatment using one-tail t test shows that t-stat value of 1.441 is not in critical region with t(27,0.05)=1.703. Therefore, the soil moisture between pre-decomposed substrate has no difference. The authors selected sugar bagasse as pre-decomposed substrate since it does not require longer process to prepare as compared with the banana stalk that needs to be finely chopped.

Fig. 8. Selection of best decomposed for the study. (left) Chopped banana stalk and (right) sugar bagasse were tested to determine the best decomposed to use. Each material were tested based on the result values from the sensor network.

3.1.1. *Soil Moisture of Baggase vs. Mean Temperature of the Day*

One way Analysis of Variance (ANOVA) has been used to find if the soil moisture of baggase differs with the mean temperature of the day. The mean at each level of test factor is of interest: the mean soil moisture at 24°C, $\mu 1$; the mean soil temperature at 25°C, $\mu 2$; and the mean soil moisture at 26°C, $\mu 3$. The factor being tested, soil temperature has three levels: 24°C, 25°C, 26°C. Since the F value is not within the critical region of the F-critical =3.12964, the results suggest that the soil temperature has no significant effect on the mean soil moisture.

3.1.2. *Soil pH Level of Pre-decomposed with Baggase*

Using the T-test one tailed test, with unknown ɑ,level of significance, $\alpha = 0.05$, and $\mu = 7$ (pH = 7, neutral), it was found out that the one-tailed test is at the left side of the t-distribution. The t statistic of -13.37609809 is within the critical region of the t-critical =1.96331077. Therefore, the pH soil level of banana stalk is less than 7.

3.2. *Individual Test Prior to System Integration*

Prior to system integration, each sensor was tested for its usability and required functionality.

3.3. *Subsystem Test prior to System Integration*

Integrating the sensors with other sensors completes the subsystem for sensor networks. These allow the coordination between subsystem. The key component of this test is the coordination of each subsystem with its sensors and needed hardware.

3.4. *System Integration Test*

Five cases are introduced for system integration test to determine the response of the system. Also, five sensors were used to determine the soil moisture and soil temperature in the worm bin, light intensity, tank water level, and ambient temperature. These values were processed by Arduino 1 and displays it through LCD at the worm bin and via Bluetooth to an android phone.

Case No.	Case Description
Case 1	Normal Condition
Case 2	Abnormal Condition
Case 3	Water level in critical condition
Case 4	Migration condition
Case 5	Power Interruption

The system is in normal condition for Case 1, soil moisture is within permissible range of 60%-80% soil moisture value. The system LED water level indicator is off and Arduino 2 is not triggered. In Case 2, under abnormal condition, soil moisture level of the worm bin is outside the accepted range, which is below 60%. Arduino 1 will send triggering signal to the sprinkler system to water the active worm bin. The sprinkler stops when the soil moisture reaches again the permissible soil moisture content of at least 60%. Water level indicator is off and Arduino 2 is not triggered. In Case 3, the tank's water level is below 50% of its full volume. The system's LED water level indicator is "on". For migration condition, under case 4, Android 2 is triggered by android 1. It then triggers the motor of the gate and lights- up the lamp. Case 5 occurs when there is a power interruption. Uninterruptible power supply (UPS) will then be triggered to provide necessary voltage for continuous process of the system.

4. Uncertainty and Predictive Analysis

The measurement of preciseness from soil moisture and soil temperature was considered between the two system to for comparative analysis. Forty five (45) sample data were taken. The authors express the level of preciseness in terms of Uncertainty Analysis in three modes: range, absolute average deviation, and relative uncertainty. The automated system gives mean soil moisture of 61.22% with $\pm$4% range versus the manual system with mean soil moisture of
60% with $\pm$ 20 %.

Measurement of preciseness through absolute average of deviation $|\Delta w|$ is:

$$|\Delta w| = \frac{1}{i^{N-1}} \sum_{i} |SM\ \ i| \tag{1}$$

The parameter SM is the soil moisture content of the vermicompost. The automatic system displays $|\Delta w|$ soil moisture of of 61.22 $\pm$ 0.641 while the manual system $|\Delta w|$ soil moisture of 60 $\pm$0.891. The relative uncertainty or the measure of preciseness of soil moisture is 2.91% for automated while 15.82% for manual system.

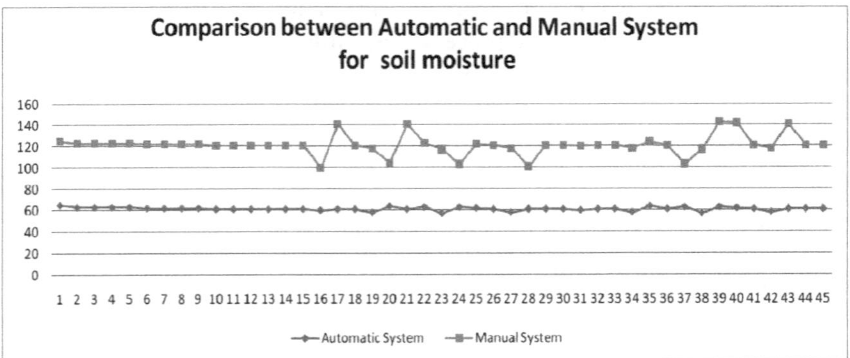

Fig. 9. Soil Moisture content using Automated and Manual System. Soil Moisture values from 45 data were compared between two systems.

The design and development of the automated system's soil moisture provides consistency of soil moisture content to the active worm bin as compared to the manual system of irrigation as seen in Figure 9.

Mean Soil temperature of the automated system gives an advantageous value as compared with the manual system.

Table 1 Mean Soil Temperature Uncertainty Analysis between Automatic Vermicomposting and Manual System.

Uncertainty Analysis	Mean Soil Temperature (Automatic)	MeanSoil Temperature (Manual)
Range	26.92°C±2.685%	25.68°C±3.5%
Absolute Average Deviation	26.92°C±0.607%	25.68°C±0.673%
Relative Uncertainty	5.348203%	10.40815%

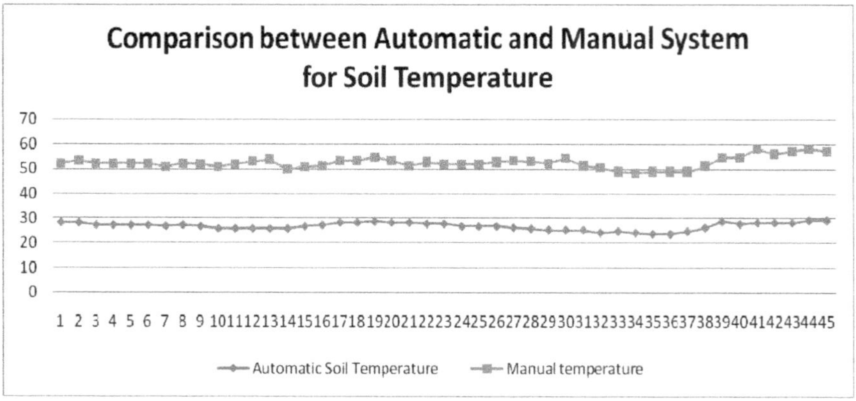

Fig. 10. Soil temperature of automated the manual system from 45 sample data.

It is evident that the measurement of preciseness of data is much more improved in automated system (Table 1). More so, the accepted temperature had been within the accepted level (Fig. 10).

4.1. *Predictive Analysis*

The predictive analysis was considered to determine moisture level for any instantaneous value temperature using curve fitting. Curve Fitting is the process of constructing a curve or mathematical function that has the best fit to a series of data points possibly subject to constraints. This is helpful if soil moisture sensor becomes faulty. Reading from soil temperature may be used to determine the soil moisture of the worm bin.

Mathematical model for the soil moisture and soil temperature:

$$y = 0.02\,x^2 - 0.98x + 73. \qquad (2)$$

The parameter y is the soil moisture and x is the soil temperature .

5. Improved Process of Automated Vermicast Production from Manual System

The nine (9) manual procedures of vermicast production were reduced to four (4) processes. Nets are no longer used to manually separate the worms from the vermicast during harvest period. Manual preparation of worm bed per layer has been improved by worm migration. Manual checking of the soil moisture has been reduced as well as the three times a day water irrigation of vermicompost.

The network sensor automatically triggers the sprinkler system when soil moisture content goes below the allowable 60% soil moisture.

The vermicast produced by the automated system was tested at the Soil Laboratory of Department of Agriculture in Region III of the country.

Result of Assay of sample of Vermicast from Automated System:

pH	5.36
Nitrogen, %	1.97
Phosphorous, %	0.80
Potassium, %	0.18
Moisture-Content, %	53.20

5.1 *Effects of the Vermicast Fertilizer to the plant Mung Beans (Vigna radiata) in Actual Application*

The mung beans were planted in three different soil mixtures: 50% vermicast – 50% normal soil; 30% vermicast – 70% normal soil; 100% normal soil. Percentages were computed on the weight of the vermicast and soil used. The total soil in one setup is 150 grams. All plants were grown for 25 days only. The presence of vermicast in the soil mixture aids in the development of the mung beans 42.6% in terms of height, 16.67% in the number of good leaves, 50% in the size of the largest leaf and 200% in the number of bean sprouts.

6. Conclusion

The paper contributions to agriculture with the integration of electronics technology greatly helps farmer to save time and effort. The automated system provides consistency of soil moisture content of the vermicompost. The two

Arduino microcontrollers were considered to avoid time delays if only one microcontroller will be used. For this project, 20 days were used and still produced acceptable nutrient for a fertilizer. Worm migration takes when the gate between two worm bins is open. It is open for 2 days to ensure total worm migration from the first worm bin towards the second worm bin. Either sugar bagasse or chopped banana stalk may be consider to use as pre-decomposed substrate, since they show similar response in terms of soil moisture and soil temperature. However, the results of vermicast production weighed 33 kilograms from a decomposed with 2 kilograms of chopped banana stalk as compared with dried sugar bagasse, the vermicast weighed 38 kilograms. Therefore, sugar bagasse was used for automated system to ensure fast production of vermicast. Data coming from sensor network are sent to an android phone via Bluetooth for monitoring and are displayed on the LCD of the worm bin.

References

1. J. Dominguez, C. A. Edwards and J. Ashby The Biology and Population Dynamics of Eudrilus eugeniae (Kinberg) (Oligochaeta) in cattle wastem solids. Vol 45 Pedobiologia, pp. 341-353 (2001).
2. E. Neuhauser, F. Kaplan, D. L. and R. Hartenstein, Life history of the earthworm Eudrillus eugeniae. Sol.16 Rev. Ecol. Biol, pp. 525-34 (1979).
3. A. Reinecke and S. Viljoen Effects of worm density on growth and cocoon production of the African Nightcrawler, Eudrilus eugeniae (Oligochaeta). Ver 29 Eur. J. Soil. Biol., pp 29– 34 (1993).
4. C. Edwards, N. Arancon and R. Sherman, "Vermiculture Technology; Earthworms, Organic Wastes, and Environmental Management, CRC Press, 2010.
5. G. Munroe, http://oacc.info/docs/vermiculture_farmersmanual_gm.pdf
6. Coulibaly, Kouassi1, Tondoh and Zoro, "Impact of the Population Size of the Earthworm Eudrilus eugeniae (Kinberg) on the Stabilization of Animal Wastes during Vermicomposting (2011).
7. M. A. Macabuhay, M.R. Abellera and A. Ticsay, "Development of an Automated Production of African Night Crawler's Vermicast with Android Application," Lecture Notes in Engineering and Computer Science: Proceedings of the World Congress on Engineering 2016, WCE 2016, 29 June–1 July, 2016, London UK, pp 363-368.
8. X. Yu, P. Wu, W. Han and Z. Zhan, " A survey on wireless sensor network infrastructure for agriculture", Computer Standards & Interfaces 35, pp. 59-64 (2013).

9. N. Singh, A. Khare, D. Bhargava and S. Bhattcharya, Optimum Moisture Requirement During Vermicomposting Using Periony Excavatus, Applied Ecology and Environmental Research, Hungary, pp. 53-62. (2004).
10. S. Gajalakshmi and S. A. Abbasi, Earthworms and Vermicomposting. Indian J. Biotechnol., pp. 486-494.
11. Ismail, Vermicology: The Bilogy of Earthworms (Orinet Longman, Hyderabad), 140 pp. (1997).
12. Guerrero III, "Earthworm Culture for Vermicompost and Vermimeal Production and for Vermiceutical Application in the Philippines — A Review", (2009).
13. Bureau of Agriculture and Fisheries Products Standards. Organic Agriculture *Philippine National Standard*, pp. 1-42 (2003).
14. Department of Agriculture. Implementing Rules and Regulations. *Organic Agriculture Act of 2010*, pp. 1-39 (2010).
15. Vermicasting (or Vermicomposting): Processing Organic Wastes through Earthworms. Onatario Ministry of Agriculture, Food and Rural Affairs. (2016), http://www.omafra.gov.on.ca.
16. Skye Instruments Limited. Soil Moisture Measurements Notes. www. skyeinstruments.com.
17. N. Mainoo, J. Whalen and S. Barrington. Earthworm Abundance Related to Soil Physicochemical and Microbial Properties in Accra, Ghanna". African J. Agricultural Res., Vol. 3. pp. 186-194 (2008).

EXTENDED RESEARCH ON PREFILTER BANDWIDTH EFFECTS IN ASYNCHRONOUS SEQUENTIAL SYMBOL SYNCHRONIZERS BASED ON PULSE COMPARISON BY POSITIVE TRANSITIONS AT QUARTER BIT RATE*

ANTONIO D. REIS†

Physics Department / U. D. Remota, University of Beira Interior - Covilha, 6200 Covilha, Portugal
adreis@ubi.pt

JOSE F. ROCHA

Electronics and Telecommunications Department / I. Telecommunications, University of Aveiro, 3800 Aveiro, Portugal
frocha@ua.pt

ATILIO S. GAMEIRO

Electronics and Telecommunications Department / I. Telecommunications, University of Aveiro, 3800 Aveiro, Portugal
amg@ua.pt

JOSE P. CARVALHO

Physics Department / U. D. Remota, University of Beira Interior - Covilha, 6200 Covilha, Portugal
pacheco@ubi.pt

This work studies the prefilter bandwidth effects in four asynchronous sequential symbol synchronizers. We choose three different bandwidths, namely $B1=\infty$, $B2=2.tx$ and $B3=1.tx$, where tx is the bit rate. First, we developed various synchronous prototypes. After, we developed various asynchronous prototypes. The synchronizer has two variants, which are the reference standard operating by both transitions at bit rate and the new proposal operating by positive transitions at quarter bit rate. Each variant has two versions, namely the manual and the automatic. The objective is to study the prefilter with the four synchronizers and to evaluate their output jitter UIRMS (Unit Interval Root Mean Square) versus input SNR (Signal Noise Ratio).

Keywords: Digital Communication Systems; Prefilter

* This work is supported by U. D. Remora
† Work partially supported by Department of Physics

1. Introduction

This work studies the prefilter bandwidth effects on the jitter-SNR behavior of four asynchronous sequential symbol synchronizers.

The prefilter, applied at synchronizer input, switches its bandwidth between three values namely, first B1=∞, after B2=2.tx and next B3=1.tx, where tx is the transmission rate.

First, we developed various synchronous prototypes, which were naturally compared, with the synchronous reference standard base. After, we developed various asynchronous prototypes, which are compared, with the asynchronous reference standard base.

We show four asynchronous synchronizers supported in two variants, one asynchronous by both transitions at bit rate with versions manual (ab-m) and automatic (ab-a) and other asynchronous by positive transitions at quarter bit rate with versions manual (ap-m/4) and automatic (ap-a/4) [1–6].

The difference between the four synchronizers is in the phase comparator. The synchronizer recoveries the clock, which is the VCO (Voltage Controlled Oscillator), that samples appropriately and retimes correctly the input data, guarantying a good quality output [7–13].

Figure 1 shows the prefilter followed of the synchronizer.

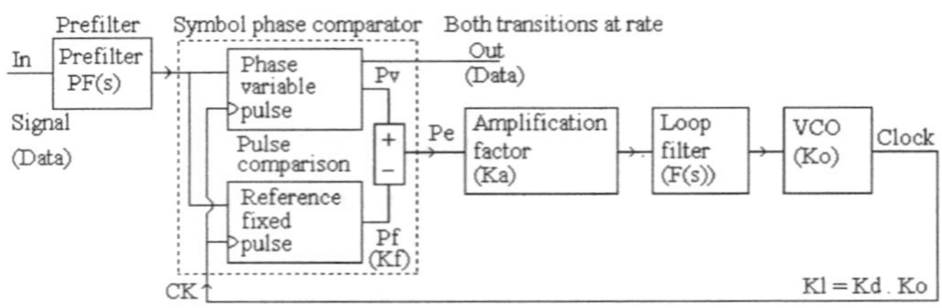

Fig. 1. Prefilter with the sequential symbol phase synchronizer.

PF(s) is the prefilter (low pass). The synchronizer has various blocks, namely, Kf is the phase detector gain, F(s) is the loop filter, Ko is the VCO gain and Ka is the loop gain factor that controls the root locus and consequently the loop characteristics.

In priori and actual-art state was developed various synchronizers, but is necessary to know their performance. This work is an extension of article [12]. The motivation of this work is to apply the prefilter to new synchronizers and to evaluate its performance with noise. Avoid some jitter resonance zones. Prevent some no linear problems. This contribution increases the know-how about synchronizers.

Following, we present the prefilter with their three different bandwidths, by decreasing order (B1=∞, B2=2.tx, B3=1.tx).

After, we present the reference standard variant, the asynchronous sequential symbol synchronizers based on pulse comparison by both transitions at bit rate, with versions manual (ab-m) and automatic (ab-a). Next, we present the new proposal variant, asynchronous sequential symbol synchronizers based on pulse comparison by positive transitions at quarter bit rate, with versions manual (ap-m/4) and automatic (ap-a/4). After, we present the project and tests. Then, we present the results. Finally, we present the conclusions.

2. Prefilter Bandwidth Effects

The prefilter is applied before the synchronizer, after, we switch its bandwidth B between three values (B1=∞, B2=2.tx, B3=1.tx), then we study the effects on the four jitter-SNR curves. Figure 2 shows the three prefilter bandwidths [12].

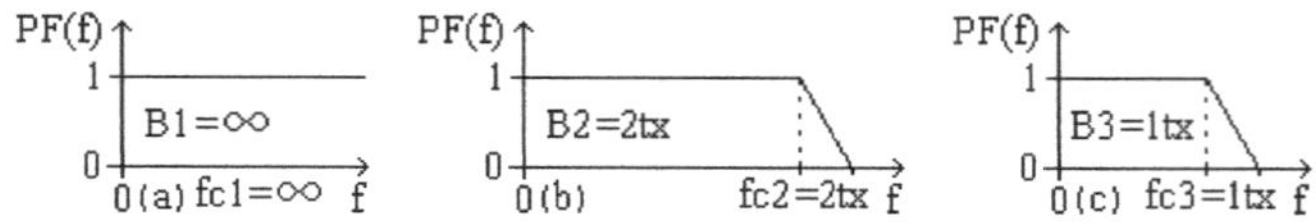

Fig. 2. Three prefilter bandwidths: a) B1=∞; b) B2=2.tx; c) B3=1.tx.

Following, we describe the prefilter (1st order Butterworth) with its three bandwidths (B1=∞, B2=2.tx, B3=1.tx).

2.1. *Prefilter with bandwidth equal infinite (B1= ∞)*

First, this prefilter (Fig. 2a) has a bandwidth equal infinite (B=∞). We will see its effects on the four asynchronous synchronizers.

2.2. *Prefilter with bandwidth equal two times the bit rate (B2= 2*tx)*

After, this prefilter (Fig. 2b) has a bandwidth equal two times the bit rate (B=2.tx). We will see its effects on the four asynchronous synchronizers.

2.3. *Prefilter with bandwidth equal one time the bit rate (B3= 1*tx)*

Next, this prefilter (Fig. 2c) has a bandwidth equal one time the bit rate (B=1.tx). We will see its effects on the four asynchronous synchronizers.

3. Reference Standard, Operating by Both Transitions at Bit Rate

The reference standard, asynchronous sequential symbol synchronizers based on pulse comparison operating by both transitions at bit rate has two versions, which are the manual (ab-m) and the automatic (ab-a). This synchronizer operates internally at same speed of bit rate [1, 2].
The versions difference is only in the phase comparator, the variable pulse Pv is common to both versions but the fixed pulse Pf is different.

3.1. *Reference, operating by both transitions at bit rate and manual (ab-m)*

The block Pv, shown below, produces a variable pulse Pv between the input bits and VCO. The manual adjustment delay with Exor produces a manual fixed pulse Pf (Fig. 3).

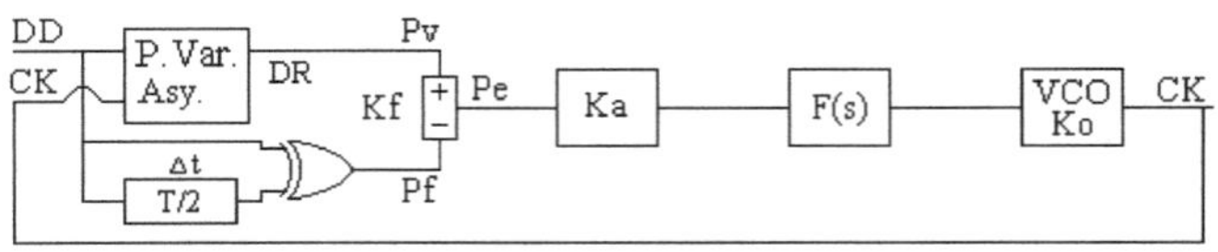

Fig. 3. Asynchronous by both at rate and manual (ab-m).

The comparison between the pulses Pv and Pf provides the error pulse Pe that forces the VCO to synchronize the input. The block Pv is an asynchronous circuit (Fig. 4).

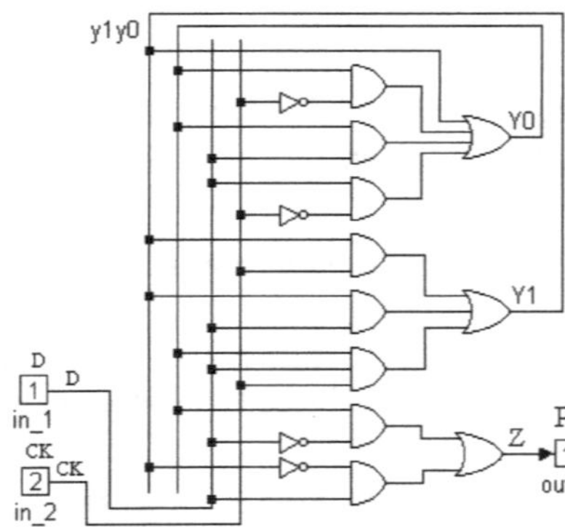

$$Y0 = y1 + y0\overline{CK} + y0D + D\overline{CK}$$
$$Y1 = y1CK + y1D + y0DCK$$
$$Z = y0\overline{D} + \overline{y1}D$$

Fig. 4. Intern aspect of the block Pv.

The error pulse Pe diminishes during the synchronization time and disappear at the equilibrium point.

3.2. *Reference, operating by both transitions at bit rate and automatic (ab-a)*

The block Pv, common with anterior, produces the variable pulse Pv between input and VCO. The block Pf, shown below, produces the comparison fixed pulse Pf (Fig. 5).

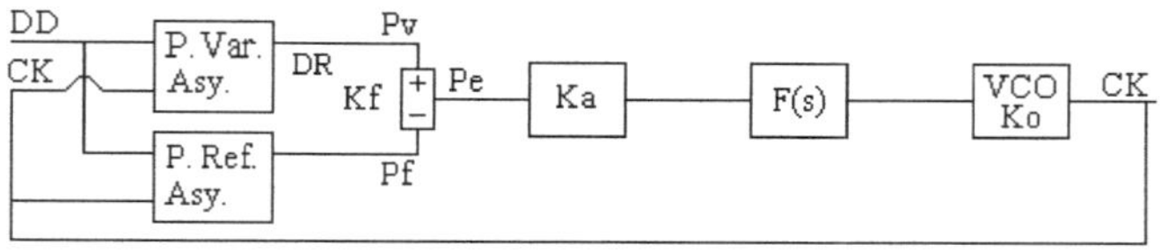

Fig. 5. Asynchronous by both at rate and automatic (ab-a).

The comparison between the pulses Pv and Pf provides the error pulse Pe that forces the VCO to follow the input. The block Pf is an asynchronous circuit (Fig. 6).

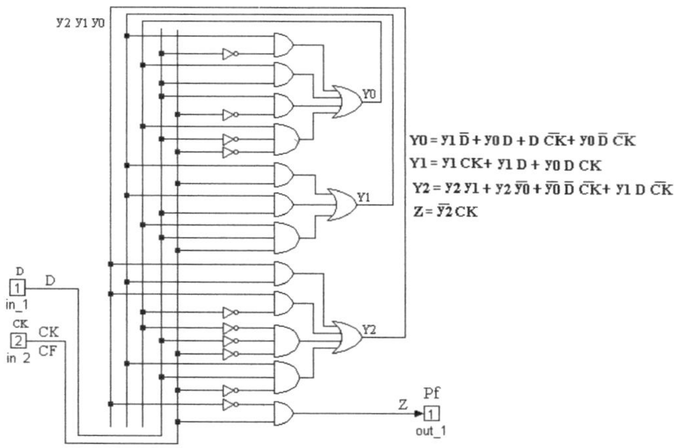

Fig. 6. Intern aspect of the block Pf.

The error pulse Pe don't disappear, but the variable area Pv is equal to the fixed area Pf at the equilibrium point.

4. New Proposal, Operating by Positive Transitions at Quarter Bit Rate

The new proposal, asynchronous sequential symbol synchronizers based on pulse comparison operating by positive transitions at quarter bit rate has also two versions namely the manual (ap-m/4) and the automatic (ap-a/4). This synchronizer operates internally at speed of only a quarter bit rate [3, 4].
The versions difference is in the phase comparator, the variable pulse Pv is common but the fixed Pf is different.

4.1. *New proposal, by positive at quarter bit rate and manual (ap-m/4)*

The block Pvp produces the variable pulse Pvp between the input transitions and the VCO. The manual adjustment delay T/2 with NOT-AND produces a fixed pulse Pfp (Fig. 7).

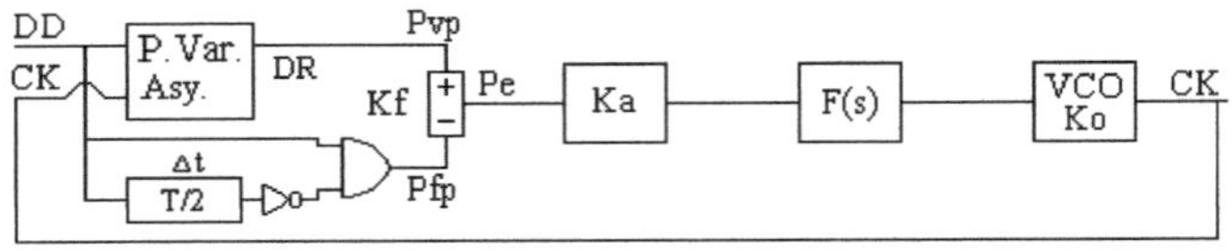

Fig. 7. Asynchronous by positive at quarter rate and manual (ap-m/4).

The comparison between pulses Pvp and Pfp provides the error pulse Pe that forces the VCO to synchronize the input. The block Pvp is an asynchronous circuit (Fig. 8).

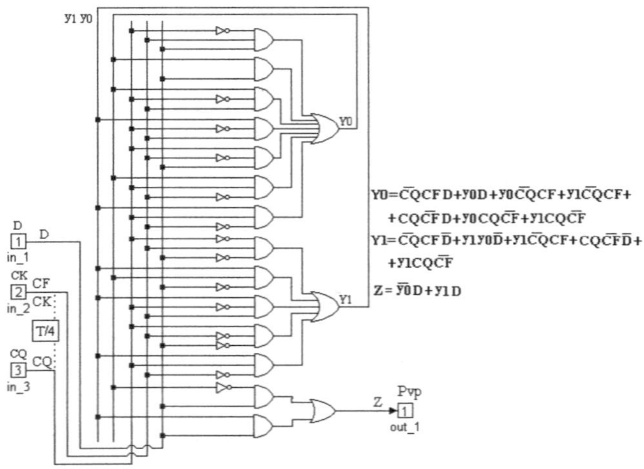

Fig. 8. Intern aspect of the block Pvp.

The error pulse Pe diminishes during the synchronization time and disappear at the equilibrium point.

4.2. *New proposal, by positive at quarter bit rate and automatic (ap-a/4)*

The block Pvp, that is common, produces the variable pulse Pvp between input and VCO. The block Pfp, shown below, produces the comparison width fixed pulse Pfp (Fig. 9).

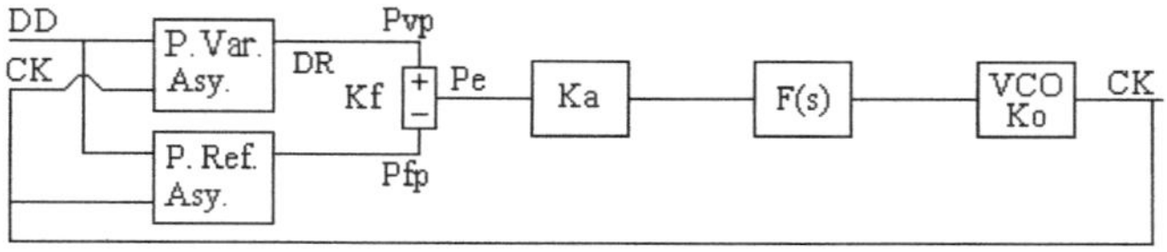

Fig. 9. Asynchronous by positive at quarter rate and automatic (ap-a/4).

The comparison between the pulses Pvp and Pfp provides the error pulse Pe that forces the VCO to follow the input. The block Pfp is an asynchronous circuit (Fig. 10).

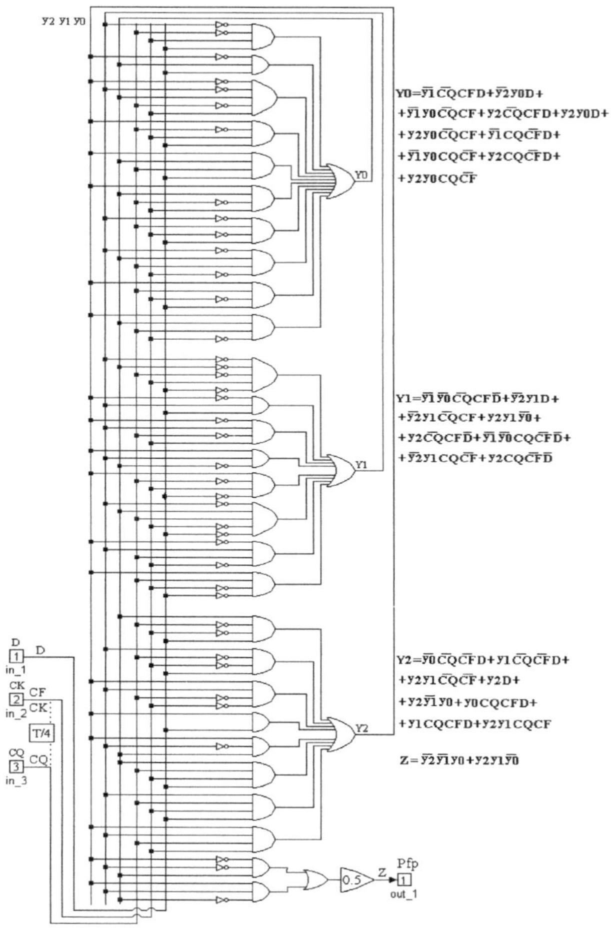

Fig. 10. Intern aspect of the block Pfp.

The error pulse Pe don't disappear at the equilibrium point, but the variable area Pvp becomes equal to the fixed Pfp.

5. Project, Tests and Results

We present the project, tests and results of the various synchronizers [5].

5.1. *Project*

We dimension all the synchronizers with the same loop gain conditions, to get guaranteed results. So, the loop gain Kl=KdKo=KaKfKo must be equal in all the synchronizers. The detector gain Kf and the VCO gain Ko are fixed, then, the loop gain value Ka controls the roots and then the loop characteristics.

For analysis facilities, we use normalized values for the bit rate tx=1baud, clock frequency fCK=1Hz, extern noise bandwidth Bn=5Hz and loop noise bandwidth Bl=0.02Hz. Then, we apply a signal power Ps= A^2_{ef} with noise power Pn= NoBn= $2\sigma n^2.\Delta\tau$.Bn, where σn is the noise standard deviation and $\Delta\tau$ =1/fSamp is the sampling period. The relation between the SNR and noise variance σn^2 is

$$SNR= A^2_{ef}/(No.Bn) = 0.5^2/(2\sigma n^2 * 10^{-3} * 5)= 25/\sigma n^2 \tag{1}$$

After, for each synchronizer, we must measure the output jitter UIRMS versus input SNR

- 1st order loop:

We use a loop filter with cutoff F(s)=0.5Hz, which is 25 times greater than Bl= 0.02Hz, what eliminates the high frequency, but maintains the loop characteristics. The transfer function is

$$H(s)=\frac{G(s)}{1+G(s)}=\frac{KdKoF(s)}{s+KdKoF(s)}=\frac{KdKo}{s+KdKo} \tag{2}$$

the loop noise bandwidth is

$$Bl=\frac{KdKo}{4}=Ka\frac{KfKo}{4}=0.02Hz \tag{3}$$

So, with ($Km=1,\ A=1/2,\ B=1/2,\ Ko=2\pi$) and loop bandwidth Bl=0.02, we obtain the Ka, respectively for some usual synchronizers, then

$$analog:\ Bl=(Ka.Kf.Ko)/4 = (Ka.Km.A.B.Ko)/4 \to Ka=0.08*2/\pi \tag{4}$$

$$hybrid:\ Bl=(Ka.Kf.Ko)/4 = (Ka.Km.A.B.Ko)/4 \to Ka=0.08*2.2/\pi \tag{5}$$

combinational: $Bl=(Ka.Kf.Ko)/4 = (Ka*1/\pi*2\pi)/4 \text{ -> } Ka=0.04$ (6)

sequential: $Bl=(Ka.Kf.Ko)/4 = (Ka*1/2\pi*2\pi)/4 \text{ -> } Ka=0.08$ (7)

For the analog PLL, the jitter is

$$\sigma_\phi^2=Bl.No/Arms^2=0.02*10^{-3}*2\sigma n^2/0.5^2=16*10^{-5}.\sigma n^2 \quad (8)$$

For the others PLLs, the jitter theoretical formula is more complicated

- 2nd order loop:

It is not considered here, but produces similar results.

5.2. *Tests*

The synchronizers are tested with the following setup (Fig. 11)

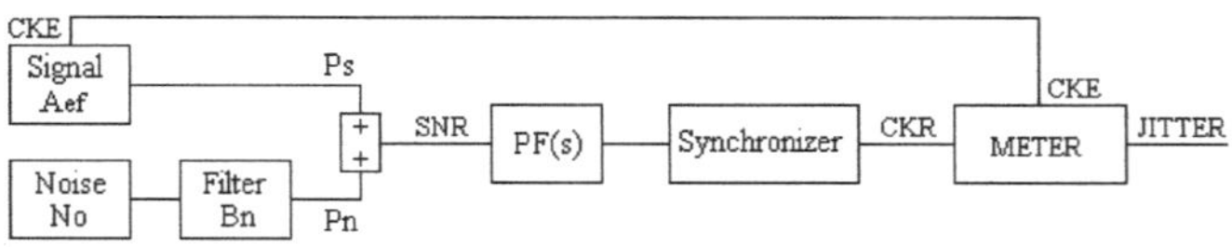

Fig. 11. Block diagram of the test setup.

The receiver recovered clock CKR with jitter is compared with the emitter original clock CKE without jitter, the difference is the jitter.

5.3. *Jitter measurer (METTER)*

The jitter measurer (METTER) consists of a RS flip-flop, an integrator and a sampler and hold (Fig. 12).

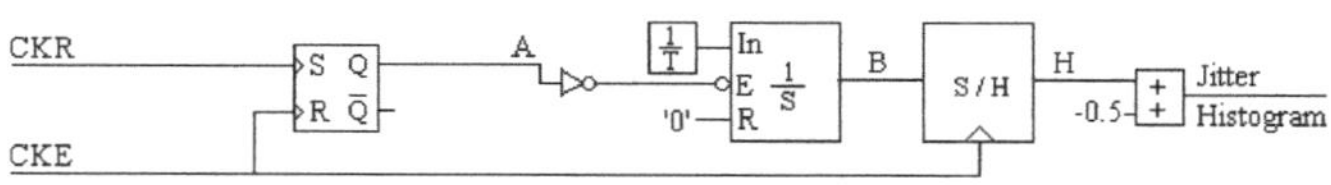

Fig. 12. The jitter measurer.

The flip flop detects the clocks relative phase variation (jitter), the other blocks convert this phase variation into amplitude variation (jitter histogram).
Following waveforms illustrate the operation mode of the jitter measurer (Fig. 13).

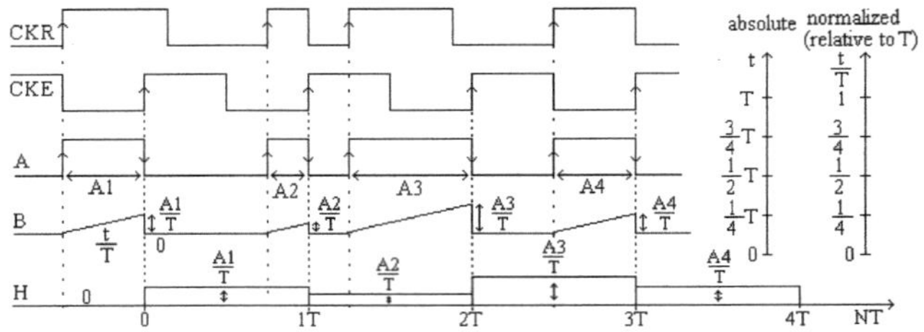

Fig. 13. Waveforms at the jitter measurer.

The jitter histogram is then sampled and processed by an appropriated program giving the jitter standard deviation in unit intervals root mean squared UIRMS.

5.4. *Results*

We will present the results, in terms of jitter - SNR, for each prefilter bandwidth with the four synchronizers.

Figure 14 shows the jitter-SNR curves for the prefilter bandwidth B1=∞ with the four synchronizers (ab-m, ab-a, ap-m/4, ap-a/4).

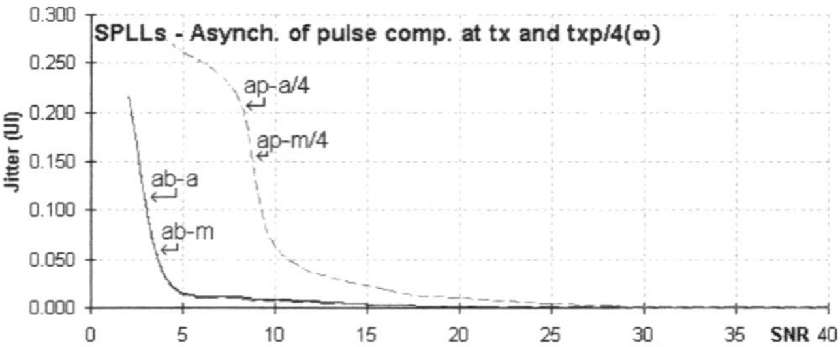

Fig. 14. Jitter-SNR curves of B1 with the four synchronizers (ab-m,ab-a,ap-m/4,ap-a/4).

For prefilter B1=∞, we verify that, for high SNR, the four synchronizer jitter-SNR curves tend to be similar. However, for low SNR, the variant asynchronous by both at bit rate with versions manual (ab-m) and automatic (ab-a) are better than the variant asynchronous by both at half bit rate with versions manual (ap-m/4) and automatic (ap-a/4).

Figure 15 shows the jitter-SNR curves for the prefilter bandwidth B2=2.tx with the four synchronizers (ab-m, ab-a, ap-m/4, ap-a/4).

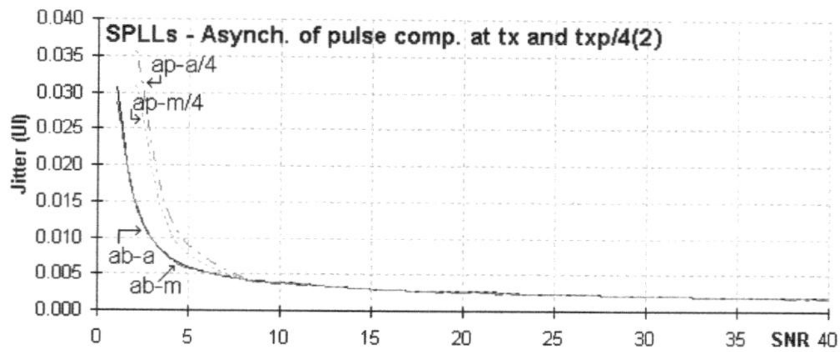

Fig. 15. Jitter-SNR curves of B2 with the four synchronizers (ab-m,ab-a,ap-m/4,ap-a/4).

For prefilter B2=2.tx, we verify that, it becomes the jitter-SNR curves more similar between themselves. For high SNR, it degrades slightly the jitter-SNR curves. However, for low SNR it benefits significantly the jitter - SNR curves.
Figure 16 shows the jitter-SNR curves for the prefilter bandwidth B3=1.tx with the four synchronizers (ab-m, ab-a, ap-m/4, ap-a/4).

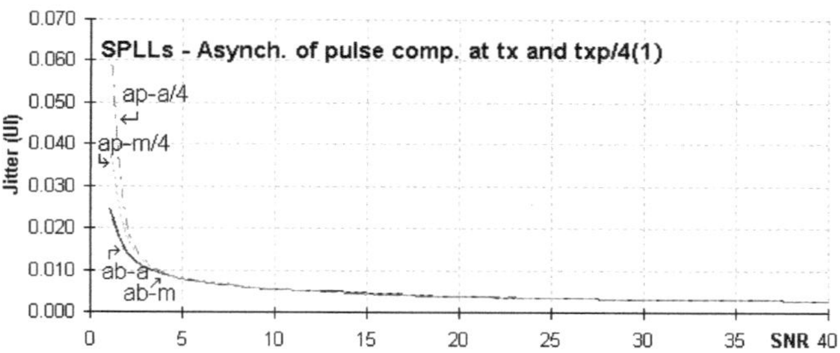

Fig. 16. Jitter-SNR curves of B3 with the four synchronizers (ab-m,ab-a,ap-m/4,ap-a/4).

For prefilter B3=1.tx, we verify that, it becomes the jitter-SNR curves still more similar between themselves. For high SNR, it degrades more the jitter-SNR curves. However, for low SNR, it benefits less the jitter-SNR curves.

6. Conclusions

We studied three prefilter bandwidths (B1=∞, B2=2.tx, B3=1.tx) with four synchronizers, one reference variant asynchronous operating by both transitions at bit rate with versions manual (ab-m) and automatic (ab-a) and other proposal variant asynchronous by positive at quarter bit rate with versions manual (ap-m/4) and automatic (ap-a/4). Then, we measured their jitter-SNR curves.
We observed that, in general, the output jitter curves decreases gradually with the input SNR increasing. There aren't jitter resonance zones, as sometimes happen with some automatic synchronous prototypes at subrates.

For prefilter B1=∞ (greater band), we verified that, for high SNR, the four synchronizers jitter curves tend to be similar, this is comprehensible since all the synchronizers are digital and have similar noise margin. However, for low SNR, the variant asynchronous by both at rate with its versions manual (ab-m) and automatic (ab-a) is better than the variant asynchronous by positive at quarter bit rate with its versions manual (ap-m/4) and automatic (ap-a/4), this is comprehensible because the variant by both transitions at bit rate has minus states than the variant by positive at quarter bit rate and then, the time to pass from the error state to the correct state is lesser in the 1st case.
For prefilter B2=2.tx (medium band), we verified that, it becomes the jitter-SNR curves more similar between themselves. For high SNR, it degrades slightly the jitter-SNR curves. However, for low SNR, it benefits significantly the jitter-SNR curves.
For prefilter B3=1.tx (lesser band), we verify that, it becomes the jitter-SNR curves still more similar between themselves. For high SNR, it degrades more the jitter-SNR curves. Also, for low SNR, it benefits less the jitter-SNR curves.
So, the prefilter, for high SNR, distorts the signal what is prejudicial, however for low SNR, it attenuates the noise what is beneficial.
So, in the new proposal, the operation by positive transitions can be advantageous in no linear systems and the operation at a quarter rate allows the system work internally at low speed and transmit externally at high speed.
In the future, we are planning to extend the present study to other types of synchronizers.

Acknowledgments

The authors are grateful to the program FCT (Foundation for sCience and Technology).

References

1. M. W. Rosenkranz, "Phase Locked Loops with limiter phase detectors in the presence of noise", *IEEE Transactions on Communications* com-30 Nº10 pp. 2297-2304. Oct. (1982).
2. J. C. Imbeaux, "Performance of the delay-line multiplier circuit for clock and carrier synchronization", *IEEE J. Selected Areas in Communications* p. 82 Jan. (1983).
3. H. H. Witte, "A simple clock extraction circuit using a self sustaining monostable multivibrat. output signal", *Electronics Letters*, Vol.19, Is.21, pp.897-898, Oct. (1983).
4. C. R. Hogge, "A self correcting clock recovery circuit", *IEEE Trans. Electron Devices* p. 2704 Dec. (1985).

5. A. D. Reis, J. F. Rocha, A. S. Gameiro and J. P. Carvalho, "A new technique to measure the jitter", *Proc. III Conf. on Telecommunications* pp.64-67 FFoz-PT 23-24 Apr. (2001).
6. M. K. Simon and W. C. Lindsey, "Tracking performance of symbol synchronizers for Manchester coded data", *IEEE Transactions on Communications* Vol. com-2.5 Nº4, pp.393-408, Apr. (1977).
7. M. J. Carruthers, D. Falconer, H. Sandler and L. Strawczynski, "Bit synchronization in the presence of co-channel interference", *Proc. Conf. on Electrical and Computer Engineering* pp. 4.1.1-4.1.7, Ottawa-CA 3-6 Sep. (1990).
8. J. Huber and W. Liu, "Data-aided synchronization of coherent CPM-receivers" *IEEE Transactions on Communications* Vol.40 Nº1, pp.178-189, Jan. (1992).
9. A. D'Amico, A. D'Andrea and Reggianni, "Efficient non-data-aided carrier and clock recovery for satellite DVB at very low SNR", *IEEE J. Sattelite Areas in Comm.* Vol. 19 Nº12 pp. 2320-2330, Dec. (2001).
10. R. Dobkin, R. Ginosar and C. P. Sotiriou, "Data synchronization issues in GALS SoCs", *Proc. 10th International Symposium on Asynchronous Circuits and Systems*, pp. CD-Ed., Crete-Greece 19-23 Apr. (2004).
11. A. D. Reis, J. F. Rocha, A. S. Gameiro and J. P. Carvalho, "Optical digital communication systems and synchronism", *Proc. 7th UBI International Conference on Engeneering (for ED)-ICEUBI 2013*, pp. CT8-11.7, Covilha-PT 27-29 Nov. (2013).
12. A. D. Reis, J. F. Rocha, A. S. Gameiro and J. P. Carvalho, "Mechanical and electromagnetic waves and the principle of the telecommunications", *Proc. Conference on ENTERprise Information Systems CENTERIS/ProjMAN/ HCIST 2015*, pp. 245-253, Vilamoura - Algarve - Portugal, 7–9 Oct. (2015).
13. A. D. Reis, J. F. Rocha, A. S. Gameiro and J. P. Carvalho, "Prefilter bandwidth effects in asynchronous sequential symbol synchronizers based on pulse comparison by positive transitions at quarter bit rate", *Lecture Notes in Engineering and Computer Science: Proc. World Congress on Engineering 2016, 'WCE 2016'*, pp. 383-386, London-UK, 29 June–1 July (2016).

SENTI-SCRM: SENTIMENT INTELLIGENCE BASED SOCIAL CUSTOMER RELATIONSHIP MANAGEMENT TOOL

AKSHI KUMAR
Department of Computer Science & Engineering
Delhi Technological University, New Delhi-110042, India
akshikumar@dce.ac.in
www.dtu.ac.in

VIKRANT DABAS
Department of Computer Science
College of Computing and Informatics
University of North Carolina, Charlotte-28223, USA
vikrantdabas@hotmail.com
www.uncc.edu

Mining business knowledge from the user-generated big social data is a challenging task and necessitates integrating intelligent models which can capture this rich brand valued knowledge across the social media platforms. This research introduces an end-to-end customer relationship management tool, Senti-SCRM (Sentiment based social customer relationship management),for brand reputation and development that automatically captures incoming posts and/or comments to the posts at the social media source (Facebook & Twitter), uncovers the post type into positive (recommendations/feedback, negative (complaint) and neutral (inquiries, interest) based on sentiment and finally directs the captured knowledge to the respective teams of business analyst & decision makers, where these can be instantly assessed, prioritized and responded.The novel Senti-SCRM tool put forward in this paper is an extension of our previous work, a Social Media Complaint Workflow Automation tool [1] which was implemented to social media posts of HDFC Bank Limited, India. The problem is similarly envisaged as a classification task which is resolved by introducing a tool based on machine learning algorithms. At the heart of this tool is the analysis of sentiment which is essentially the key to knowledge discovery from the available textual data. It is implemented to social media posts of Samsung Mobile Facebook page & a comparative performance analysis for same has been performed.

Keywords: Customer Relationship Management; Sentiment Analysis; Social Media.

1. Introduction

Recent years have transformed Web into a social ecosystem which probes relationships and conversations between customers and organizations facilitating a multi-channel ubiquitous customer management system. Social Customer

Relationship Management (SCRM) is defined as the customer relationship management fostered by engaging customers through social networking sites, such as Twitter, Facebook, YouTube etc. It is extending the capabilities of the Social Web technologies for a strategic, profitable customer relationship management (CRM) within organizations. Mining knowledge from the user-generated big social data is a challenging task and necessitates integrating intelligent models within the conventional business models. Brand Management [2] has emerged as the primary areas of study & practice, where extracting knowledge from opinion-rich costumer sourced social media is valuable to automate, monitor, analyze & manage businesses."Word-of-Mouth" [3] is vital and strategic determinant of customer satisfaction and brand management (positive/negative word-of-mouth) can harness this knowledge by incorporating sentiment intelligence within the business model, thus setting up a novel model of Sentiment intelligence based social customer relationship management (Senti-SCRM).

More formally, Brand management is the science of crafting and sustaining a brand [4]. It implies defining, positioning and delivering the brand by constantly managing its tangible and intangible elements. Tangibles associated with product brands include the product itself, its packaging, pricing, availability, etc. With service brands, tangibles include customers' experience. The intangibles comprise of emotional connections and expectations with products and services [5]. Further, the rise of the user-friendly social media technologies has given a smart control in the hands of the customers making him conversant, vocally empowered & omnipresent in the current business model. Pertinent studies have reported that approximately 1.2 billion people worldwide use Facebook [6] to follow brands (53 % at least once per month), learn more about brands (65 %) or hear of others' experiences with brands (70%) [7]. YouTube [8] is another technology which has been used as a brand publicity medium. Users like, react, comment on, or share this brand information making it as a word-of-mouth which can be either negative or positive. Similarly, conversations on Twitter [9] can be acquired to understand the public response to a newly launched brand and at the same time determine the brand market divide by analyzing competitors' presence. To capture this rich brand valued knowledge across the social media platforms, the organizations need to introduce & embed a social CRM for firstly identifying *who's* talking about your brand and the relevant related keywords online, *what* they are talking about and *where* they are saying it [10]. We propose sentiment classification is the core capability which can guide this application of social CRM in businesses for analyzing and monitoring public opinion endorsing brand management, and more specifically, brand reputation management.

Our research identified key areas where this alliance of social media to the brand can directly be linked to customers. Managing customer relations will require text mining the available big user-generated data to support delivery of targeted types of information to analysts and decision makers within the domains of inquiry. As a business intelligence tool, Senti-SCRM will help gaining insights to:

- Product/ service/policy launch & monitoring for both public & beta versions without demographic restrictions. It is about investigating the real-time market data, conversations, customer experiences, and competitor's impact for brand engagement.
- Inquiries & Expression-of-Interest from consumers for promoting business sales.
- Complaints, negative feedbacks for improved responsiveness& reputation management.

This research introduces an end-to-end customer relationship management tool for brand reputation & management that automatically captures incoming posts and/or comments to the posts at the social media source (Facebook/ Twitter/YouTube), uncovers the post type into positive (recommendations), negative (complaint) or neutral (inquiries, interest, generic) based on sentiment and finally directs the captured knowledge to the teams of business analyst & decision makers, where these can be instantly assessed, prioritized and responded. The problem is envisaged as a classification task which is resolved by introducing a tool based on machine learning algorithms. At the heart of this tool is the analysis of sentiment which is essentially the key to knowledge discovery from the available textual data. Sentiment analysis or Opinion mining, as it is sometimes called, is one of many areas of computational studies that deal with opinion oriented natural language processing [11]. It typically involves the classification of text into categories such as "positive", "negative" and in some cases "neutral".

The novel Senti-SCRM tool is an extension of our previous work, a Social Media Complaint Workflow Automation tool [1] which was implemented to social media posts of HDFC Bank Limited, India & tested for various classification models. The tool automated the process that manages complaints at a social media source by firstly capturing incoming posts and/or comments to the posts, extracting the complaints by uncovering the post-type based on sentiment and finally directing them to the concerned departments of the bank for a quick response strategy. Senti-SCRM, based on the same idea to yardstick public mood toward products &services, at regular intervals, say quarterly, consists of two primary modules, namely the Brand Reputation module which subsumes the previously evaluated complaint management along with newly integrated Brand Development module for an organic, powerful brand development. The intent is to utilize sentiment intelligence with a clear motivation to pro-actively manage the destructive viral power of social media for brand reputation.

The paper is organized as follows: Section 2 gives a brief explanation of related background work in this direction. Section 3 describes the workflow model of the proposed Senti-Social Customer Relationship Management tool. Section 4 gives a brief overview of the implementation and analysis using a case study of Samsung Mobile social media sources. Section 5 concludes the research undertaken while discussing the potential applications & directions of future work.

2. Related Work

The provision of customer service via social media channels has become axiomatic, especially due to its cheap, convenient and public nature. Having a dedicated Social Media Manager, who keeps checking the media platforms every ten minutes, may seem a solution but in a B2C organizations, manually keeping track, monitoring and analyzing the velocity and variety of data available in a time-bound manner is tedious & inefficient. Though many social media monitoring tools like Hootsuite [12], TweetReach [13], Social Mention [14], Twazzup [15], Addictomatic [16], amongst others, are available in market, they fail to provide a complete end-to-end customer relationship management model and require dedicated social media business analytics team to study the reports generated by them. The work presented in this research is a completely automated model to manage customer relations at the social media source using intelligent text mining for a responsive & controlled market presence. The next section describes the proposed workflow tool.

3. The Proposed Sentiment Intelligence based Social Customer Relationship Management Tool (Senti-SCRM)

Conventional Customer Relationship Management (CRM) systems are like a single-channel pipeline process supported by dedicated phone lines & support e-mail ids, primarily restricted demographically. Radical changes can be foreseen by harnessing the participative & collaborative Social Web. Based on this, the proposed Sentiment intelligence based Social Customer Relationship Management Tool (Senti-SCRM) primarily works as an online brand management tool at the social media. It is perceived as a two-level classification paradigm that uses sentiment analysis on the social media posts and/or comments to comprehend social interactive dynamics mapping positive posts (including comments) to business analytics, negative posts (complaints) to brand reputation management & neutral posts (sales, inquiries, interest) to brand management. Finally this polarity classified posts are further directed to the respective customer service teams to automate a complete workflow for guiding, monitoring and managing customers, thus providing absolute CRM. For both levels of classification, Naïve Bayes Classifier, its 3 variants, namely Multinomial Naïve Bayes, Bernoulli Naïve Bayes and Gaussian Naïve Bayes & Support Vector Machine are employed. The tool is implemented to the social media posts of Samsung Mobile (Facebook) [17] & tested for various classification models. Figure 1 illustrates the basic workflow architecture of the proposed tool.

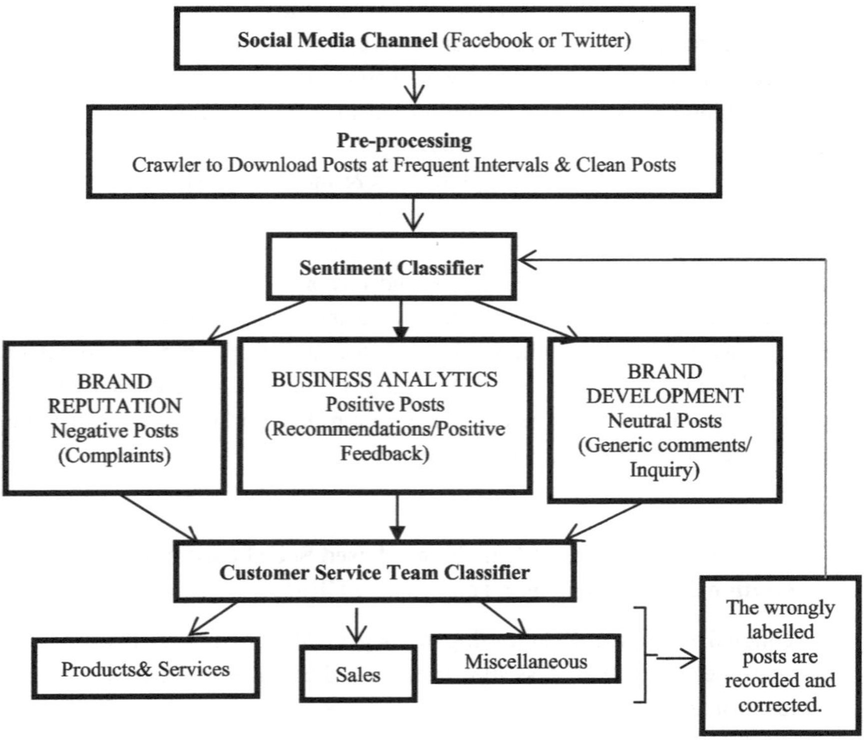

Fig 1. The senti-SCRM tool.

The following sub-sections expound the details pertaining to each component of the proposed workflow:

3.1. ***Pre-processing***

Large publically available social media datasets are crawled to fetch data using explicit APIs. Data extracted in its original form is hard to use for analysis directly. It is often incomplete, noisy & inconsistent [18]. Thus, the pre-processing consists of the following two steps:

3.1.1. *Crawling social media channels to capture posts*

According to the Internet Society's "Global Internet Report 2015" [19], there are more than 3 billion people online globally who use Internet for communicating, sharing, finding information, publishing & commerce amongst others. Statistics also reveal that more than 2 billion people worldwide have social accounts (Facebook, Twitter etc.). The Social networking websites like Facebook and Twitter which are a substantial, unparalleled source of user-generated content are easy to crawl and extracting online posts from them can be done in a straight forward manner. To get large publically available Facebook datasets, we use Facebook's graph API. We form a URL & hit the graph API to receive JSON object responses containing data about Likes, talking about etc. Required data is extracted from these JSON objects and saved in a separate file for further processing. For implementation & analysis both Facebook posts and comments have been considered.

3.1.2. *Cleaning Module*

Data cleaning for quality is significantly desirable for higher accuracy & better results. The data extracted from Facebook contains many undesirable symbols, words, URLs, E-mail ids, etc. that may slow down the applied algorithms or even make them perform incorrectly. Hence, it becomes necessary to incorporate a data cleaning module. The following strategy has been identified to clean the downloaded data for processing:

- Removal of # tags: Facebook & Twitter have incorporated hashtags, which works on personal profile posts, fan page posts, group posts, event posts, and all comments.
- Removal of "@": @ symbol *is* used for tagging or to "mention" someone on Facebook or Twitter.
- Removal of URLs & E-mail ids.

- Removal of numbers, digits, alphanumeric data, punctuations, symbols.
- Expanding short-forms (acronyms), if any.
- Removal of non-English posts by encoding them into ASCII and if encoding produces any exception, then reject them, otherwise, select the posts.
- Removal of stop-words.

The original posts/comments/ & the cleaned posts comments/ are kept temporarily at each step. The cleaned post is used for processing & the original posts are stored for further use.

3.2. *Classification of posts*

Cleaned posts and/or commentsto the post are labeled and trained over the first classifier (Sentiment Classifier). These are classified into three categories, namely:

- *Negative Posts (Complaints)*: an expression of dissatisfaction on a consumer's behalf to a responsible party.
- *Neutral Posts (Generic post/Inquiry)*:text with no clear sentiment polarity. An NLP based analysis is further done on all neutral posts to identify interrogative neutrals (neutral posts which end with a question mark) as we observed that these correspond to request for information from the concerned the organization such has brand inquiries, product details & sales.
- *Positive Posts (Recommendation/Positive reviews)*: posts that are basically feedback from customers that show a positive response or a set of posts that are either for advertisement purposes or are completely irrelevant. These can further be analyzed for real-time brand and business evaluation & projection.

Thus, the sentiment classifier extracts the posts/comments to categorize them into positive, neutral and negative classes which are forwarded to the customer service team classifier. This classifier categorizes the results from previous classifier to allocate them to respective team to provide end-to-end customer relationship management. The customer service team class files which are created are explained as follows:-

- *Products & Services:* This team handles product complaints & service issues thereby supporting the customers in an empathetic & responsive manner thus enabling brand reputation management.

- *Sales:* This is the team which supports brand development by offering inquiry assistance looking out specifically for neutrally classified interrogative posts.
- *Miscellaneous:* Positive comments/feedback, generic posts are maintained for further business analytics.

The sentiment and customer service team classifiers are trained over a labeled data set. Once the posts have been labeled and separated as positive, negative & neutral posts, the original posts that were mapped to the cleaned posts during pre-processing are sent to the respective teams for further action & proceedings and the cleaned posts are discarded. In case, they reach a wrong team, i.e., if it is found to be incorrectly classified, the posts are labeled/marked with the correct team and are used to update the classifiers.

For both classifiers, the learning algorithms considered were Naïve Bayes (NB), its 3 variants, namely, Multinomial Naïve Bayes (MNB), Bernoulli Naïve Bayes (BNB) and Gaussian Naïve Bayes (GNB) & Support Vector Machine (SVM) [20, 21]. Naïve Bayes classifier uses a bag-of-words [22] approach for sentiment analysis on the content. The Bernoulli Naive Bayes can be used when in our problem the absence of a particular word matters. For example Bernoulli is commonly used in Spam or advertisement detection with very good results.
The Multinomial Naive Bayes is suitable for classification with discrete features & is typically used when occurrence of the word matters more than the frequency. For example it does not really matter how many times someone mentions the words "bad", "awful", "terrible", but rather only the fact that he does. A Binarized Multinomial Naive Bayes is used for the sentiment classification (first classifier), i.e., it has binary weighting function where the value 1 means that the word occurs in the particular document, and 0 means that the word does not occur in this document. A tf-idf Multinomial Naive Bayes is used for the team classification (second classifier). The tf-idf approach assumes that the importance of a word is inversely proportional to how often it occurs across all documents. The Gaussian Naïve Bayes & support vector machine were also used for a comparative performance analysis.

4. Implementation and Performance Analysis

To clearly illustrate the effectiveness of the proposed method, a case study of Samsung Mobile Facebook page is presented. Figure 2 shows the Samsung Mobile Facebook page with some sample posts.

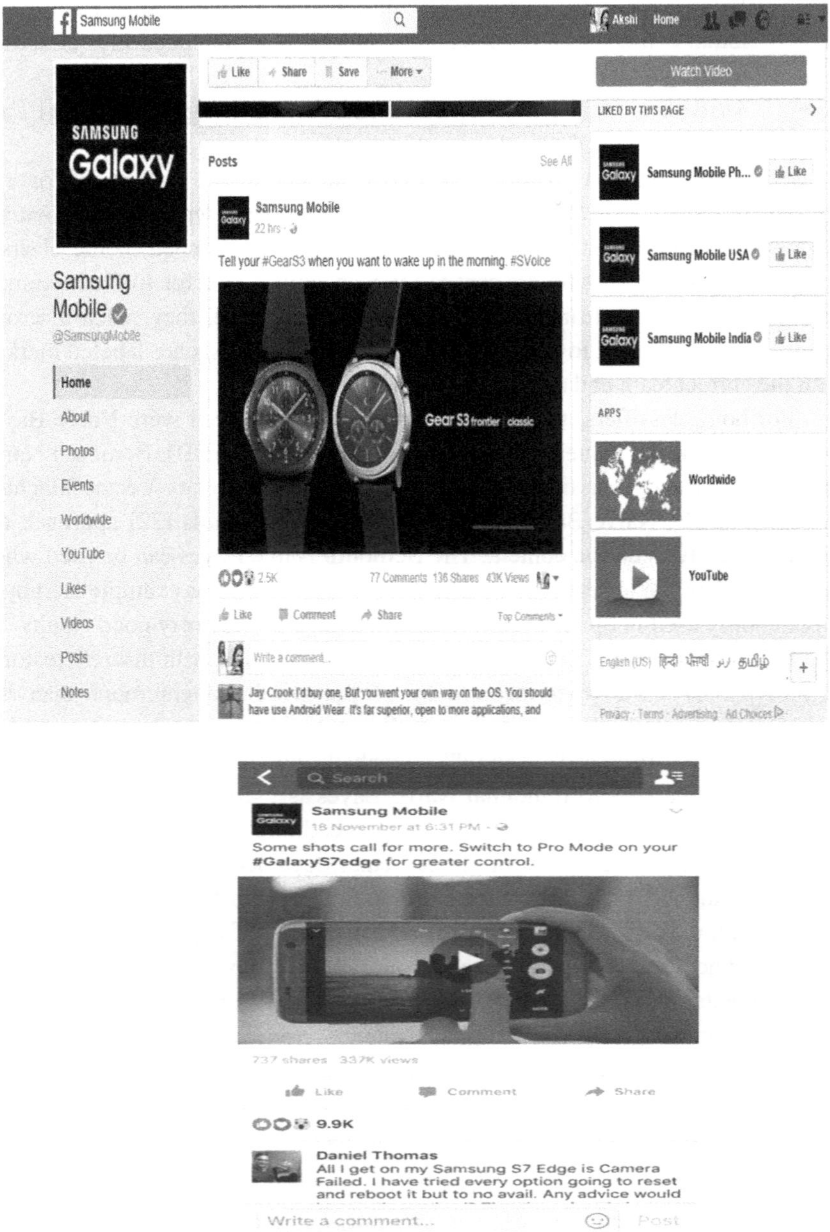

Fig. 2. Snapshot of Samsung Mobile Facebook page with some sample posts.

Each of the components of the workflow is exemplified to demonstrate how it actually works by firstly identifying the complaints from the Facebook post & then referring it to the appropriate team which can cater a quick response, thus automating the entire process in an efficient manner.

4.1. *Pre-processing*

4.1.1. *Crawling social media channels to capture posts*

To download the Facebook data, an access token was generated by registering as a developer on Facebook. The next step was to make an HTTP GET request to the Facebook graph API using the urllib.request module in python. As a response, a JSON dump of the data was received and useful information (user who posted, post text, time of posting) was extracted & stored in a CSV file.

4.1.2. *Cleaning*

The cleaning process as discussed in the previous section, involved removing hashtags like #SamsungMobile etc., tagging symbol @ (@Samsung_Mobile), URL supporting additional information (links to company's website) and any e-mail ids that are included in the post, example: support@samsungmobile.com. It also scrubs numbers (mobile numbers/dates etc.) and very common stop-words like "a, an, the, this, that".

4.2. *Classification*

As discussed in the previous section, for both levels of classification, Naïve Bayes Classifier (NB), its 3 variants, namely Multinomial Naïve Bayes (MNB), Bernoulli Naïve Bayes (BNB) and Gaussian Naïve Bayes (GNB)& Support Vector Machine (SVM) have been employed. The results from the first classification task for finding sentiment from the post are shown below in Table 1.

Table 1. Sample Results of Sentiment Classifier.

Posts	Sentiment	Class
My LGG5 has manual settings on the camera and you can take a picture in Raw image. The camera on the LGG5 is amazing. All though the LGG5 doesn't have the premium feel as the S7 and S7 edge I like the fact you can remove the battery in the LGG5 and the performance of this phone is amazing and never lets me down.	Positive	Recommendation

Really amazing gadget. I just love it!	Positive	Recommendation
All I get on my Samsung S7 Edge is Camera Failed. I have tried every option going to reset and reboot it but to no avail. Any advice would be greatly received? The phone is only just over a month old.	Negative	Complaint
Purchased a samsung new mobile on 3-3-2015. 6 times they repaired- Finally they told in Dec15 it is beyond repair, agreed for refund, but was not getting- follow up-finally went to consumer court. Now I got refund and penalty imposed on them by court. Samsung is not dependable- cheating company.	Negative	Complaint
My Gear S broke like all the Gear S. To fix it Samsung ask me to pay almost the same amount to buy a new one!!! Shame!!! Samsung made a mistake in the project and now costumers has to pay?!?!?!?!?!? I always bought Samsung but from now stop I can't trust anymore!!!!!!! Shame shame shame!!!	Negative	Complaint
hey Samsung is the s6 will receive the 7.1 nougat update?	Neutral	Generic Posts/Inquiry
Just tell me when the new Samsung Note is coming our already?	Neutral	Generic Posts/Inquiry
Need 7.0 for J5	Neutral	Generic Posts/Inquiry
Samsung Mobile hello I have samsung galaxy s4. I just need to ask 1 thing that my mobile keeps going on and off sometimes while battery remains full? Can you tell me what is the problem?	Neutral	Generic Posts/Inquiry

The customer service team classifier categorizes the posts acquired into various teams for further necessary relationship management. A snapshot for identified neutral posts which are sent to the Sales team is given in Figure 3.

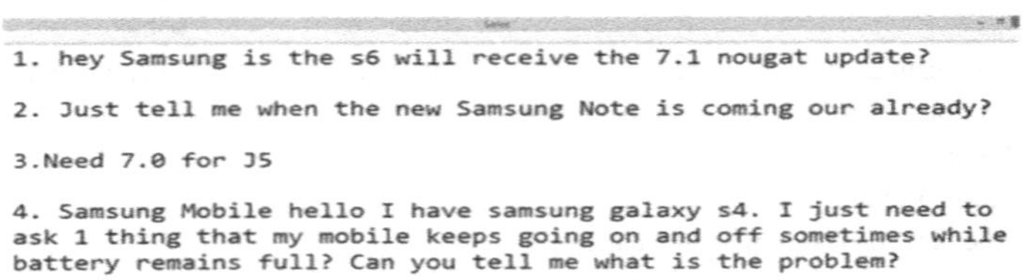
1. hey Samsung is the s6 will receive the 7.1 nougat update?

2. Just tell me when the new Samsung Note is coming our already?

3.Need 7.0 for J5

4. Samsung Mobile hello I have samsung galaxy s4. I just need to ask 1 thing that my mobile keeps going on and off sometimes while battery remains full? Can you tell me what is the problem?

Fig. 3. Snapshot of result from team classifier with inquiries fitting to the Sales team.

4.3. *Results & Performance Analysis*

This section discusses the result & does a performance analysis for both the classification tasks with all learning algorithms exercised.

Table 2 shows the performance analysis of the sentiment classifier(first classifier) on the basis of accuracy and time. The Multinomial Naive Bayes Classifier gave the highest accuracy. Bar graphs are shown for the same in Figures 4 and 5 respectively.

Table 2. Analysis of sentiment classifier.

Algorithm	Accuracy (%)	Time(sec)
Naive Bayes Classifier	78.672	0.1562
Multinomial Naive Bayes Classifier	83.658	0.0312
Bernoulli Naive Bayes Classifier	79.662	0.0468
Gaussian Naive Bayes Classifier	67.662	0.0156
Support Vector Machines	76.331	0.0312

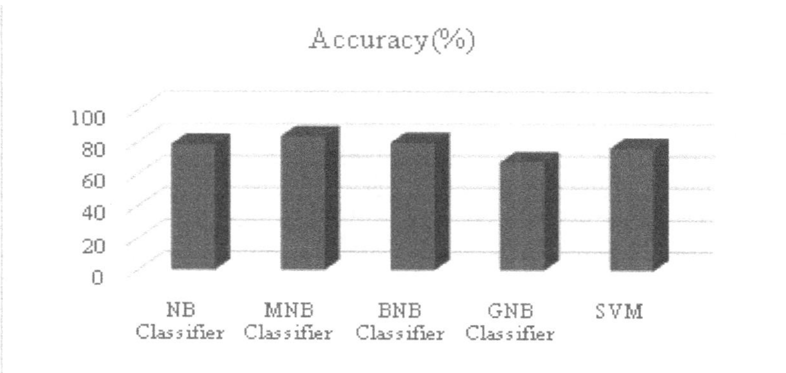

Fig. 4. Accuracy analysis of sentiment classifier.

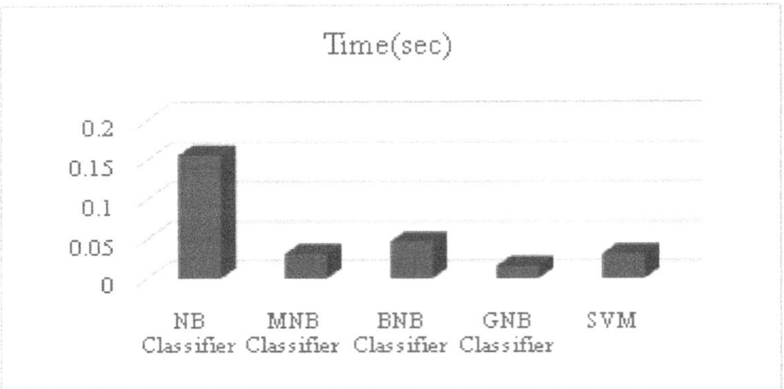

Fig. 5. Time analysis of sentiment classifier.

For the team classifier (second classifier), Table 3 shows its performance on the basis of accuracy and time. Here too, the Multinomial Naive Bayes Classifier gave the highest accuracy and Gaussian Naïve Bayes gave the lowest accuracy. Gaussian Naïve Bayes works well for continuous numerical data and owing to our discrete dataset, its performance was compromised. Bar graphs are shown for the same in Figures 6 and 7 respectively.

Table 3. Analysis of team classifier.

Algorithm	Accuracy (%)	Time(sec)
Naive Bayes Classifier	70.670	0.2035
Multinomial Naive Bayes Classifier	74.672	0.1002
Bernoulli Naive Bayes Classifier	73.667	0.0989
Gaussian Naive Bayes Classifier	61.669	0.0311
Support Vector Machines	69.331	0.0899

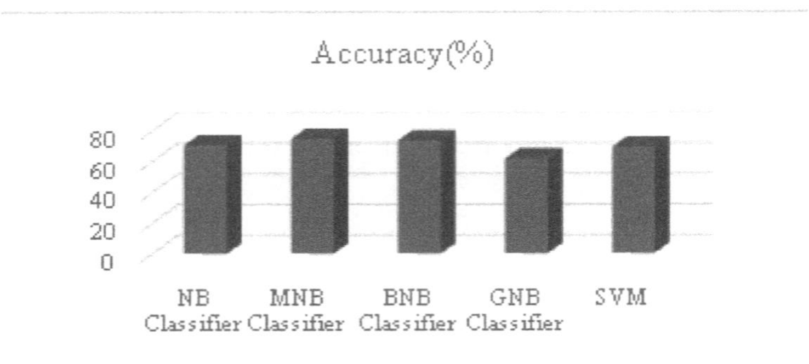

Fig. 6. Accuracy analysis of team classifier.

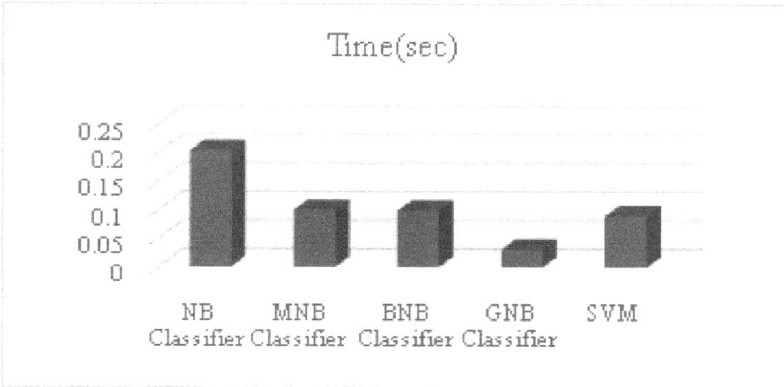

Fig. 7. Time analysis of team classifier.

5. Conclusion

Over the past decade organizations have been compelled to examine new business models & technological innovations to match the tech-savvy, socially mindful Gen-Y. Thus tracking social media for word-of-mouth (feedback/ recommendations/complaints) should be regarded as a pro-active response policy. The tool proposed in this paper is an automation of the process that manages customer relationship at a social media source by firstly capturing incoming posts and/or comments to the posts, extracting the recommendations, inquiries &complaints by uncovering the post-type based on sentiment and finally directing them to the concerned customer service teams of the organization where they can be instantly assessed, prioritized and responded. A comparative performance analysis for implementation on Samsung Mobile Facebook page posts gave encouraging results with a highest accuracy of approximately 83% for sentiment classification & approximately 75% for team classification. To further improve the accuracy results & optimize results in terms of response time, we intend to examine association mining and/or co-information metric for relation analysis amongst co-occurring terms. Also currently we examine the interrogative neutral posts as inquiries using basic natural language processing, alternative techniques to identify & improve the accuracy in this category. A fine-grain sentiment analysis, probably emotion-based is envisioned too. Moreover, due to the complexity of the project, we currently have only implemented and tested it for Facebook whereas exploring Brand management using Twitter & YouTube can be beneficial too and help defining a robust business model.

References

1. A. Kumar and Dabas, V, "*A Social Media Complaint Workflow Automation Tool using Sentiment Intelligence*", Lecture Notes in Engineering and Computer Science: Proceedings of the World Congress on Engineering 2016, WCE 2016, 29 June - 1 July, 2016, London, UK, pp. 176-181.
2. K. Keller and D. R. Lehmann, *"Brands and Branding: Research Findings and Future Priorities,"* Marketing Science, 25(6), 2016, pp. 740-759.
3. F. Buttle, "*Word-of-Mouth: Understanding and Managing Referral Marketing,"* Journal of Strategic Marketing, 6, 1998, pp. 241-54.
4. *Blog: http://www.revenew.com/blog/the-importance-of-brand-management*
5. Management Study Website on Brand Management: *https://www.managementstudyguide.com/brand-management.htm*
6. Facebook Homepage: *www.facebook.com*
7. Olaf Maecker, Christian Barrot, Jan U. Becker, *"The effect of social media interactions on customer relationship management."* Business Research, April 2016, Volume 9, Issue 1, pp. 133–155.
8. YouTube Homepage: *www.youtube.com*
9. Twitter Homepage: *www.twitter.com*
10. Blogposton How brands must evolve from CRM to Social CRM: *https://www.brandwatch.com/blog/how-brands-must-evolve-from-crm-to-social-crm/*
11. A. Kumar and T. M. Sebastian, "*Sentiment analysis: A perspective on its past, present and future"*, International Journal of Intelligent Systems and Applications (IJISA) 2012, 4(10), pp 1-14.
12. Hootsuite Homepage: www.Hootsuite.com
13. TweetReach Homepage: www.TweetReach.com
14. SocialMention Homepage: www.SocialMention.com
15. Twazzup Homepage: www.Twazzup.com
16. Addictomatic Homepage: www.Addictomatic.com
17. Samsung Mobile Facebook Page: https://www.facebook.com/Samsung Mobile
18. G. J. Williams and S. J. Simoff, "*Data Mining: Theory, Methodology, Techniques and Applications*", Book published by Springer, 2007.
19. Internet Society Global Report: http://www.internetsociety.org/global internetreport/2015/assets/download/IS_web.pdf

20. R. Schapire, *"Machine Learning Algorithms for Classification",* Princeton University, Handouts, 2011.
21. F. Sebastiani, *"Machine learning in automated text categorization"*, ACM Computing Surveys (CSUR),Vol. 34, Issue 1, March 2002, pp. 1–4.
22. J. Rennie, L. Shih, J. Teevan and D. Karger, *"Tackling the Poor Assumptions of Naïve Bayes Text Classifiers",* Proceedings of ICLM, 2003, pp. 616–623.

THE ROLE OF TABLET PC TOWARDS STUDENTS' LEARNING IN HIGHER EDUCATION

H. K. YAU*, Y. F. LEUNG and C. Y. YEUNG†
Department of Systems Engineering and Engineering Management,
City University of Hong Kong,
Kowloon Tong, Kowloon, Hong Kong
**honkyau@cityu.edu.hk*
www.cityu.edu.hk

The purpose of this study was to examine the effectiveness of the use of tablet PCs for students' learning in higher education and students' perspective towards using tablet PCs for learning. University students' perceived learning and perceived engagement towards the use of tablet PCs in learning, and perceived engagement towards perceived learning. Questionnaire-based survey was conducted and 187 sample data were collected from students in different universities in Hong Kong. The background information of respondents and their perspective regarding the use of tablet PCs in learning were obtained for data analysis, followed by result discussion.

Keywords: Having own tablet PC; Perceived Learning; Perceived Engagement

1. Introduction

Over the past decade, mobile technologies have been developing rapidly and play an important role in tertiary education. The incorporation of mobile technologies into education can stimulate students' interest in learning by potentially changing the dynamics of learning environment as well as providing a liberalization of learning, based on the mobile devices' ability of being connected to people, ideas and information of students' own choosing [1]. The utilization of mobile devices allows high flexibility and convenience for students to access the learning materials anywhere, anytime, so there is a growing expectation of the ability to "work, learn, and study whenever and wherever they want" [2]. Yau and Leung [23] have worked out the gender difference of self-efficacy and attitudes towards the use of technology in learning in Hong Kong higher education. However, it seems that no one have done how the use of tablet PCs can influence students; perceived learning and engagement. Therefore, this paper fills this research gap. The purpose of this study is to analyze the question: How does the use of tablet PCs influence students' perceived learning and engagement?

2. Literature Review

According to Kahveci [3], experience is associated with students' familiarity of using the technology. Students' experience in using a technology is one of the determinants of their perceptions in the use of the technology in learning [4]. Past studies have reported that students with more experience in using the technology would have more positive perceptions [4–7] as well as higher engagement of using the technology in learning [8, 9].

In the context of this study, it is assumed that students who have their own tablet PCs are more experienced in using tablet PCs. Therefore, based on the influences of students' experience of the use of technology, we hypothesize that:

H1: Students having their own tablet PC are more positive in perceived learning than those who do not have a tablet PC.

H2: Students with their own tablet PC have higher perceived engagement than those who do not have a tablet PC.

3. Methodology

The questionnaire for this research was designed based on the studies from "iLearning: The Future of Higher Education? Student Perceptions on Learning with Mobile Tablets" by Jonathan P. Rossing, Willie M. Miller, Amanda K. Cecil and Suzan E. Stamper [10], as well as "Exploring College Students' Attitudes and Self-Efficacy of Mobile Learning" by Shih-hsien Yang [11]. These two literatures are related to the impact of mobile technologies including tablets on students' learning and students' perceptions on the use of tablets in learning, which are the critical elements of this study. The questions (Table 1) in the main parts are rated on a five-point Likert scale ranging from strongly disagree (=1) to strongly agree (=5).

Table 1. Items of questionnaire.

Question	Item	Factor Loading
1	Using a tablet PC helps me apply course content to solve problems	0. 651
2	Using a tablet PC helps me learn the course content	0. 680
3	Using a tablet PC helps me connect ideas in new ways	0. 711
4	Using a tablet PC helps me participate in the course activity in ways that enhanced my learning	0. 750

5	Using a tablet PC helps me develop confidence in the subject area	0. 703
6	Using a tablet PC helps me develop skills that apply to my academic, career and/or professional life	0. 729
7	The use of tablet PCs motivates me to learn the course material more than class activities that do not use tablet PCs	0. 647
8	I participate more in class with the use of tablet PCs than during activities that do not use tablet PCs	0. 669
9	My attention to the task(s) is greater using a tablet PC	0. 605
10	A tablet PC is more convenient compared to a desktop or laptop computer	0. 548
11	It is easier to work in a group using a tablet PC than in other group activities	0. 671
12	A tablet PC is an important supplement to the class	0. 646

After the questionnaire had been constructed, pilot study was done before distributing the questionnaire in a large scale. According to Lowe [12], piloting the questions on a small group of people could effectively save valuable time as any problems would be revealed at this stage so that the quality of the questionnaire can be improved. In the pilot study, ten university students were invited to complete the questionnaire without any explanation from the researcher about it so as to find out whether they could understand the questions or not. The ten participants were then asked individually for feedback about the questionnaire after completing it for the pilot test. After receiving some feedback from the participants, the order of Part 2 and Part 3 in the questionnaire was exchanged as putting the issue about self-efficacy before the one about attitudes towards the use of tablets in learning would make the questionnaire better organized. Meanwhile, questions 18 to 20 are modified to a way such that they can become positive statements. Besides, for the question about CGPA in Part 1, an option of "N/A" was added since some respondents may not have their GPA released yet. Moreover, some wording was also modified to make the questions easier to understand.

The finalized questionnaire could be then distributed to the target group which is university students studying in Hong Kong. The questionnaires were distributed during the lectures in one of Hong Kong university so that a large number of questionnaires could be collected. For respondents from other universities, questionnaires were distributed through the help of my friends in different universities. However, not all distributed questionnaires were returned.

Totally, 300 questionnaires were distributed physically and 213 copies were returned. However, only 187 copies were useful and 26 copies were incomplete or with invalid data. The successful response rate was:
= collected and valid questionnaires / total number of questionnaires distributed
= 187/300 × 100%
= 62.3%

Since the questionnaires collected were just raw data at this stage, the raw data would be input into the software "Statistical Package for the Social Science" (SPSS) in order to facilitate the process of data analysis [13]. The data would be analysed and meaning information with useful statistics would be generated by SPSS.

The statistical tool used in this research was "Statistical Package for the Social Science" (SPSS). This software, which is used by many scientists and professionals, is user-friendly and sophisticated for statistical analysis, data management and data documentation and able to generate everything we need for data analysis [14].

Factor analysis would be used to identify groups of similar items, create and validate scales, ensuring that the items are associated with scales as well as to reduce the number of variables for further analysis. Anderson and Gerbing [15] stated that the items in a questionnaire are validated in unidimension and then undergo factor loading as well as corrected item-total corrections to test for their values and internal validity. Fornell and Larcker [16] proposed that the acceptable value for factor loading and the corrected item-total corrections is 0.3 for verifying items. Hence, if the factor loading or corrected item-total corrections was lower than 0.3, the item should be eliminated.

The reliability of scales would be evaluated by Cronbach's alpha, which is a common measure of internal consistency to see how closely related the items are [17]. Reliability is defined as the fact that a scale should consistently reflect the construct it is measuring [18]. All the scales measured should be higher than the suggested cut-off criteria, i.e. Cronbach's alpha > 0.7 ([19], such that they could be said to be reliable. However, Cronbach's alpha is sensitive to the number of items in the scale, so the resulted values would be likely to be lower than the acceptable value (0.7) if the scale only contained a few number of items. Hence, Briggs and Cheek [20] proposed a more appropriate way which is to report the mean inter-item correlation for the items with an optimal range of 0.2 to 0.4.

After the validation of measurement scales, the differences between the means of two independent groups would be tested by independent t-test [21], while the differences between the means of more than two groups would be tested by one-way analysis of variance (ANOVA). Besides, the Pearson product-moment correlation coefficient (r) would be used to determine whether there is a relationship between two factors [22].

Independent Samples T-test is used to find out whether two groups are statistically different. By using SPSS, two groups of data can be formed by grouping the selected variables in order to generate their mean difference as well as other statistics. The t-test consists of two parts which are "Levene's Test for Equality of Variances" and "t-test for Equality of Means".

"Levene's Test for Equality of Variances" is used to check whether the variances of two tested groups are the same and determine the suitable t-value to be used. For significance value of Levene's Test greater than 0.05, it indicates equal variance for the two groups so the data of "equal variances assumed" will be used for both groups. Otherwise, the data of "equal variances not assumed" will be used.

After determining the variances assumed for the two groups, "t-test for Equality of Means" is used to test any significant difference exists between them. For 2-tailed significance value equal to or less than 0.05, it implies that significant difference exists in the means of the two groups. Otherwise, there is no significant difference between the two groups.

4. Results

In this research, questionnaires were collected from 187 respondents. Among all respondents, about 43.3% are male and 56.7% are female. The major age range of the participants is between 18 and 23, which accounts for about 88.2%, while there are 7.5% of participants aged from 24 to 30 and the other age ranges only accounts for about 4% in total.

Besides, regarding their educational level, 49.2% of respondents are Year 3 students, 25.7% of them are Year 2, and 17.6% of them are postgraduate students. Both Year 1 and Year 4 students account for 3.2% and the others account for 1.1% of respondents. Most of the participants are engineering students, which account for 64.7%. There are 4.3% of respondents majoring in business and 3.7% of them majoring in science. Participants majoring in social science and creative media account for 2.7% and 1.1% respectively. 23.5% of respondents are studying in other majors such as nursing.

Moreover, the majority of respondents are full-time students, which account for 90.9%, while part-time and exchange students account for about 8% and 1.1% respectively. Most of the participants have GPA with the range of second honour in which 47.1% are 3.0-3.49 and 29.9% are 2.5-2.99. There are about 8.0% of respondents whose GPA is in the range of first honour and about 9.1% have GPA in the range of third honour.

In addition, 65.2% of respondents have their own tablet PCs while 34.8% of them do not have.

Summary of Students Having Tablet PC on Perceived Learning and Summary of Students Having Tablet PC on Perceived Engagement was shown in Tables 2 and 3 respectively.

Table 2. Means and standard deviation.

	Having a tablet PC	N	Mean	Std. Deviation
Perceived learning	Yes	122	3.6202	0. 72956
	No	65	3.2538	0. 65159

The significance value in Levene's Test is 0.315, which is greater than 0.05, so the data in the row of "Equal variances assumed" should be used. The value of significance (2-tailed) in "t-test for Equality of Means" is 0.001, which is smaller than 0.05, indicating a significant difference between students having own tablet PC and those without own tablet regarding perceived learning.

According to Table 2, the mean value for students with own tablet PC is 3.6202, while the mean for students without own tablet PC is 3.2538, which is lower than that for students having own tablet PC. This indicates that students with own tablet PC have more positive perception in learning with the use of tablet than those who do not own a tablet.

Based on the result of the independent t-test ($p=0.001<0.05$), the hypothesis H1 was supported.

H1: Students having their own tablet PC are more positive in perceived learning than those who do not have a tablet PC.

Table 3. Means and standard deviation.

	Having a tablet PC	N	Mean	Std. Deviation
Perceived engagement	Yes	122	3.4645	0. 75583
	No	65	3.1103	0. 57303

The significance value in Levene's Test is 0.070, which is greater than 0.05, so the data in the row of "Equal variances assumed" should be used. The 2-tailed significance value in "t-test for Equality of Means" is 0.001, which is smaller than 0.05. Hence, there is a significant difference between students having own tablet PC and those without own tablet regarding perceived engagement.

According to Table 3, the mean value for students with own tablet PC is 3.4645, while the mean for students without own tablet PC is 3.1103, which is lower than that for students having own tablet PC. This shows that students with own tablet PC have higher perceived engagement with the use of tablet than those who do not own a tablet.

Based on the result of the independent t-test ($p=0.001<0.05$), the hypothesis H2 was proved.

H2: Students with their own tablet PC have higher perceived engagement than those who do not have a tablet PC.

5. Discussion

The results reveal that students with own tablet PC have more positive perceived learning as well as higher perceived engagement in the use of tablet PCs for learning than those who do not have their own tablet PCs. It implies that students with more experience in using tablet PCs generally have more positive perceptions and higher engagement on learning with tablet PCs than those with less experience.

The finding is supported in other researches which show that students who are more experienced in using a technology would have more positive perceptions and higher engagement of using the technology in learning than those with less experience [4–7]. This is because students with more previous experience are more familiar with the functionalities of tablet PCs and thus they have more confidence in using tablets. This in turn promotes their perceived learning and engagement on learning with the use of tablets.

□

6. Conclusion

This study has evaluated the role of tablet PC towards students' learning in higher education by identifying students' perceived learning and perceived engagement towards using tablet PCs for learning. Meanwhile, the effects of some personal characteristics and the relationships among these factors of student perspective have been investigated as well. Questionnaire was designed and distributed to university students in order to collect relevant data from respondents. The collected data were analysed and statistical results were generated by Statistical Package for the Social Science (SPSS).

The main limitation of this study is the uneven sample size of respondents. In the survey, the majority of respondents are engineering students and most of the respondents are studying bachelor degree. Since the learning behaviour and environment of students with different majors and different education levels can be vary a lot, the data findings in the research are limited to represent students in higher education.

On the other hand, the majority of respondents are from the University the researcher were studying due to the convenience for distribution of questionnaires. Thus, the distribution of the sample size is not even enough and not representable to reflect the general opinions of students in higher education. Random sample should be used instead of a convenient sample in order to achieve more a representative study.

This project mainly just focused on the student perspective regarding their perceived learning and perceived engagement towards using tablet PCs for learning but no concrete findings on how well the use of tablets can enhance students' learning outcomes. Thus, future studies can investigate other factors such as the effects of the use of tablets on students' performance. Besides, future research can involve teacher perspective as well in order to examine the all-round effectiveness of integrating tablet PCs into educational setting.

Moreover, this study only examined the effects of gender and experience on the use of tablet PCs so future research can investigate the effects of other personal characteristics. In addition, as tablet PC is identified as an educational tool that can be applied for students with different learning styles, future studies can also examine the effects on different learning styles.

References

1. J. Traxler, Learning in a mobile age. *International Journal of Mobile and Blended Learning*, 1, 1, 1-12 (2009).
2. L. Johnson, S. Adams and K. Haywood, The NMC Horizon Report: 2011 K-12 Edition. Austin, TX: The New Media Consortium (2011), http:// media.nmc.org/iTunesU/HR-K12/2011/2011-Horizon-Report-K12.pdf

3. M. Kahveci, Students' perceptions to use technology for learning: measurement integrity of the modified Fennema-Sherman attitudes scales. *Turkish Online Journal of Educational Technology*, 9, 1, 185-201 (2010).
4. I. Vekiri and A. Chronaki, Gender issues in technology use: Perceived social support, computer self-efficacy and value beliefs, and computer use beyond school. *Computers & Education*, 51, 3, 1392-1404 (2008).
5. B.H. Loyd and C. Gressard, The effects of sex, age, and computer experience on computer attitudes. *AEDS journal*, 18, 2, 67-77 (1984).
6. C. Comber, A. Colley, D.J. Hargreaves and L. Dorn, The effects of age, gender and computer experience upon computer attitudes. *Educational Research*, 39, 2, 123-133 (1997).
7. V. Venkatesh, M. Morris, G. Davis and F.D. Davis, User acceptance of information technology: Toward a unified view. *MIS Quarterly*, 27, 3, 425-478 (2003).
8. T.F.N. Laird and G.D. Kuh, Student experiences with information technology and their relationship to other aspects of student engagement. *Research in Higher Education*, 46, 2, 211-233 (2005).
9. E.M. Mercier, B. Barron and K.M. O'Connor, Images of self and others as computer users: The role of gender and experience. *Journal of Computer Assisted Learning*, 22, 5, 335 (2006).
10. J.P. Rossing, W.M. Miller, A.K. Cecil and S.E. Stamper, iLearning: The future of higher education? Student perceptions on learning with mobile tablets. *Journal of the Scholarship of Teaching and Learning*, 12, 2, 1-26 (2012).
11. S.H. Yang, exploring college students' attitudes and self-efficacy of mobile learning. *Turkish Online Journal of Educational Technology*, 11, 4, 148-154 (2012).
12. M. Lowe, *Beginning Research: A Guide for Foundation Degree Students*. (Routledge, 2006).
13. J.R. Fraenkel and N.E. Wallen, *How to Design and Evaluate Research in Education* (*5 ed.*). (McGraw-Hill, New York, 2003).
14. D. Muijis, *Doing Quantitative Research in Education with SPSS*. SAGE Publications Inc., USA, 2004).
15. J. Anderson and D. Gerbing, Structural modelling in practice: A review and recommended two-steps approach. *Psychological Bulletin*, 10, 3, 411-423, (1988).
16. C. Fornell and D. Larcker, Structural equation models with unobservable variables and measurement error: Algebra and statistics. *Journal of Marketing Research*. 15, 282-388 (1981).

17. J.C. Nunnally, *Psychometic Theory,* (McGraw-Hill, New York, 1978).
18. A. Field, *Discovering statistics using SPSS,* (Sage publications, 2009).
19. C. Fornell and D. Larcker, Structural equation models with unobser vable variables and measurement error: Algebra and statistics. *Journal of Marketing Research*. 15, 282-388 (1981).
20. S.R. Briggs and J.M. Cheek, The role of factor analysis in the development and evaluation of personality scales. *Journal of personality*, 54, 1, 106-148, (1986).
21. G. Hole, Using SPSS to perform t-tests. *Research skills*, 1-10 (2009).
22. R. Ho, *Handbook of univariate and multivariate data analysis and interpretation with SPSS,* (CRC Press, 2006).
23. H. K. Yau and Y. F. Leung, Gender difference of self-efficacy and attitudes towards the use of technology in learning in Hong Kong Higher Education, Lecture Notes in Engineering and Computer Science Proceedings of the International MultiConference of Engineers and Computer Scientists 2016, IMECS 2016, 16-18 March, Hong Kong, pp. 819-821.

THE LATEST DEVELOPMENTS IN VIRTUAL COMPUTER-INTEGRATED MANUFACTURING SYSTEMS*

SON DUY DAO†, KAZEM ABHARY and ROMEO MARIAN
School of Engineering, University of South Australia, Mawson Lakes Campus, Mawson Lakes, SA 5095, Australia
†son.dao@mymail.unisa.edu.au

Virtual Computer-Integrated Manufacturing (VCIM) is a new manufacturing system, developed to help small and medium-size enterprises survive and thrive in today's competitive global market. VCIM systems are still, conceptually, being developed. In this chapter, the latest developments in VCIM systems are presented. There are two important developments in the systems, i.e. VCIM production scheduling model and VCIM production scheduling optimization approach. Current progress in the scheduling model as well as the optimization approach is presented herein.

Keywords: Collaborative Shipment Scheduling, Genetic Algorithm, Virtual Computer-Integrated Manufacturing, VCIM Production Scheduling Model.

1. Introduction

Computer-Integrated Manufacturing (CIM) is a manufacturing system in which computers are used to control the production processes [1]. Virtual Enterprise (VE) is a temporary alliance of enterprises, working together to share skills, competencies and resources to better respond to business opportunities [2]. To inherit the good characteristics and addresses the limitations of CIM and VE, a new concept called Virtual Computer-Integrated Manufacturing (VCIM) was proposed. VCIM is a network of manufacturing enterprises, in which the members temporarily work together in an integrated manner to create a temporary production system to fulfil the customer order. This temporary production system and temporary co-operation will dissolve after the order is fulfilled [3]. It should be noted that a new temporary production system will be created for the next customer order, by a group of the same and/or different enterprise members. In addition, manufacturing enterprises in a VCIM system are locally and/or globally distributed.

* Work supported by Australian Government in form of Endeavour Award.

Generally speaking, VCIM is a combination of CIM and VE concepts, which was first proposed by Lin [4] in 1997. The main purpose of a VCIM system is to effectively exploit the distributed manufacturing resources. VCIM is a pretty new concept and it is still being developed. As can be seen from the definition above, VCIM is a global manufacturing system which is very integrated and flexible. The integration and flexibility help a VCIM system exploit the distributed manufacturing resources effectively in today's global market. Of course, a lot of work needs to be done to make VCIM systems a reality, e.g. information sharing, security, copyright, membership, quality control, etc. Although some work has been done [5-13], a real VCIM system does not exist yet. Nevertheless, inspired by the big success of Alibaba model [14, 15], the researchers strongly believe that VCIM system will be a promising solution for many small and medium-size manufacturing enterprises in near future.

To date, all research related to VCIM systems focus on developing VCIM production scheduling model and optimization. Production scheduling in VCIM systems is defined as (1) selecting manufacturing enterprises and (2) scheduling the related shipments to create a temporary production system to fulfil the given customer order [16]. Production scheduling is very important to VCIM systems because it will affect the quality, cost as well as lead time of products. In addition, VCIM production scheduling is a complex optimization problem, which requires an advanced optimization method. In this chapter, the latest developments in scheduling model as well as optimization approach for VCIM production scheduling problem are presented.

2. The Latest Production Scheduling Model for VCIM Systems

There have been a number of research works developing the production scheduling model for VCIM systems [3, 5-8, 10-13]. The latest VCIM production scheduling model in the literature to date is a model published by [16], in which multiple customer orders, collaborative shipment scheduling as well as uncertainties are simultaneously considered. The model can be described as follows.

Consider:

- A VCIM system has a number of assembly agents as well as component suppliers, which are both globally and locally distributed.
- The VCIM system is capable of providing a number of products for customers worldwide and each product is made by assembling a number of components.

- Each component supplier can provide certain components for certain products.
- Each assembly agent can do final assembly for certain products.
- In the current planning horizon, several customer orders with the same and/or different products are being requested.

Determine:

- Allocation of the required components of the requested products to the component suppliers?
- Allocation of the requested products to the assembly agents?
- The related component and product shipping schedules?

So that:

A temporary VCIM production system can be formed to simultaneously fulfil all of the customer orders with (1) minimum total cost, (2) minimum average completion time and (3) maximum average reliability, while all of the given constraints are satisfied.

Conditions:

- Cost and release time of each product component provided by different component suppliers are different from one another but known in advance.
- All product components can be independently produced.
- Assembly cost and time of each product assembled in different assembly agents are also different but given in advance.
- Shipping cost and time between any two locations are different from each other but known in advance.
- All component release times, product assembly times and shipping times are stochastic parameters with known distribution functions, mean values as well as standard deviations.
- One product is assembled in one assembly agent only.
- One assembly agent is capable of assembling a certain number of products.
- All component suppliers and assembly agents have enough resources such as labor, material, etc. to perform their designed functions 24 hours a day, 7 days a week.

The VCIM production scheduling model is illustrated in Fig. 1 with a typical production scheduling solution to fulfil 8 customer orders (C1, C2, …, C8). The system has 14 assembly agents (A1, A2, …, A14) and 20 component suppliers (S1, S2, …, S20), distributed worldwide. In the illustrated solution, component suppliers S2, S5, S6, S9 and assembly agent A13 were selected to serve customers C1-C3; component suppliers S7, S18, S20 and assembly agent A14 were selected to fulfil the customer orders C4-C6; and component suppliers S10, S16, S20 and assembly agent A2 were selected to serve customers C7-C8. Two kinds of shipments, separate shipment and collaborative shipment, were used in the solution as shown in Fig. 1. It is noted that the separate shipment herein refers to a shipment carrying the component(s) produced by only one component supplier or a shipment carrying finished product(s) to only one customer. In contrast, the collaborative shipment means a shipment carrying the components made by more than one component supplier or a shipment carrying finished products to more than one customer. Examples of separate and collaborative shipments are shown in Fig. 1.

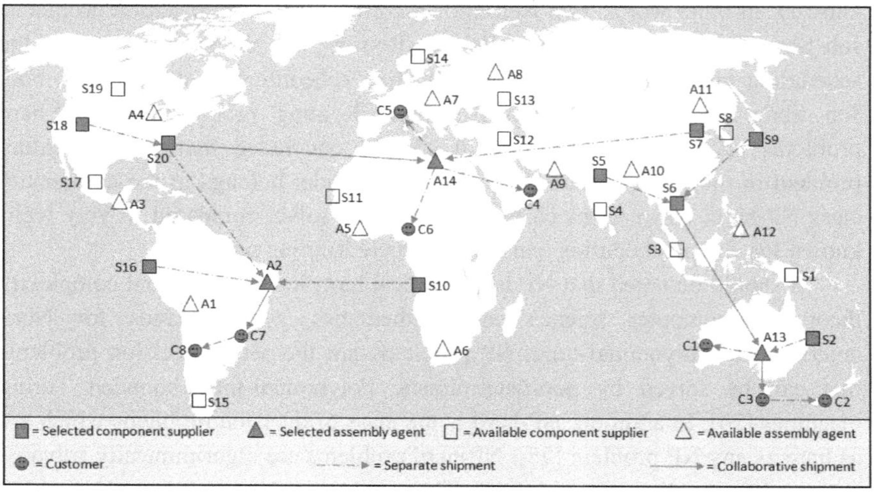

Fig. 1. The latest production scheduling model for VCIM systems [16].

As can be seen from the description above, some general ideas about the VCIM production scheduling model have been expressed. However, details of the model, i.e. mathematical version of the model, has not been presented yet. The authors stated that the mathematical version of the model will be released in near future.

3. The Latest Optimization Approach for VCIM Production Scheduling Problem

Combinatorial optimization is a class of computational problems dealing with the questions of selecting the optimum from a finite set of possibilities [17]. Obviously, the problem in the VCIM production scheduling model is a combinatorial optimization problem. Finding the global optimal solutions to combinatorial optimization problems is not a trivial problem. Although it is easy to deal with small-size problems, it is very hard, if not impossible, when coping with large-size or real-world problems [18].

From computational complexity theory point of view, the scheduling problem in the model is a NP-hard problem, because of the following reasons. First, this scheduling problem would become a Job Shop Scheduling problem [19, 20] if the collaborative shipment is not considered. Second, the scheduling problem would become an Open Vehicle Routing problem [21, 22] if the initial problem would be simplified when the partners, i.e. the selected component suppliers and assembly agents, are fixed, and no shipping route with branches is allowed. In other words, two well-known combinatorial optimization problems, Job Shop Scheduling and Open Vehicle Routing, are the special cases of the scheduling problem in the VCIM production scheduling model. Finally, both Job Shop Scheduling and Open Vehicle Routing problems are NP-hard problems [19, 21-23]. Therefore, it can be concluded that the scheduling problem in the VCIM production scheduling model belongs to the complexity class NP-hard, and it is much more difficult to solve, compared to very well-known Job Shop Scheduling and Open Vehicle Routing problems.

It should be noted that NP is a technical term in computational complexity theory in computer science and mathematics, which stands for Non-deterministic Polynomial-time. NP problems are the set of decision problems that can be solved by non-deterministic polynomial-time bounded Turing machines [24]. In addition, NP-hard is the class of decision problems which are as hard as any NP problem [25]. NP-hard problems are algorithmically solvable but computationally intractable [25]. There is no exact method that can find the global optimal solutions to NP-hard problems in polynomial time, and fast approximate heuristics and meta-heuristics are the popular approaches to search for high-quality/practical solutions [26].

Genetic Algorithm (GA) is one of the most popular meta-heuristics [27], often used to solve complex large-scale optimization problems. GA is capable of handling large search spaces, continuous/discrete problems and complex constraints [28]. GA is the best candidate for solving the optimization problem in the VCIM production scheduling model. The authors are developing a robust GA for this class of optimization problems, in which an innovative chromosome encoding, modified crossover/mutation operators, an innovative algorithm structure, a systematic parameter tuning method and an effective strategy for including Monte Carlo simulation in the optimization algorithm are proposed. Details of the GA will be published in near future.

4. Conclusion

In this chapter, the latest developments in VCIM production scheduling model as well as the related optimization approach have been presented. In the latest model, multiple customer orders, collaborative shipment scheduling and uncertainties are simultaneously considered. This is the most comprehensive VCIM production scheduling model in the literature. To solve the optimization problem in this model, a GA with an innovative algorithm structure, a systematic parameter tuning method and an effective strategy for including Monte Carlo simulation in the optimization algorithm, is being developed.

References

1. Miller, F.P., A.F. Vandome, and J. McBrewster, *Computer-integrated manufacturing,* 2010, Mauritius: VDM Publishing House.
2. Camarinha-Matos, L.M. and H. Afsarmanesh, *The virtual enterprise concept,* in *Infrastructures for virtual enterprises – Networking industrial enterprises,* L.M. Camarinha-Matos and H. Afsarmanesh, Editors. 1999, Kluwer Academic Publishers: The Netherlands. pp. 3-14.
3. Zhou, N., *et al., Development of an agent based VCIM resource scheduling process for small and medium enterprises,* in *International MultiConference of Engineers and Computer Scientists,* 2010: Hong kong. pp. 39-44.
4. Lin, G.C.I., *The latest research trends in CIM,* in *The Fourth International Conference on Computer Integrated Manufacturing*1997: Singapore. pp. 26-33.
5. Wang, D., *The development of an agent-based architecture for virtual CIM,* 2007, University of South Australia, Adelaide: Adelaide.
6. Wang, D., S.V. Nagalingam, and G.C.I. Lin, *Development of an agent-based virtual CIM architecture for small to medium manufacturers.* Robotics and Computer-Integrated Manufacturing, 2007. **23**(1): pp. 1-16.
7. Wang, D., S.V. Nagalingam, and G.C.I. Lin, *Development of a parallel processing multi-agent architecture for a virtual CIM system.* International Journal of Production Research, 2004. **42**(17): pp. 3765-3785.
8. Nagalingam, S.V., G.C.I. Lin, and D. Wang, *Resource scheduling for a virtual CIM system,* in *Process Planning and Scheduling for Distributed Manufacturing,* L. Wang and W. Shen, Editors. 2007, Springer London. pp. 269-294.
9. Lin, G.C.I., S.V. Nagalingam, and J. Zhou, *Virtual CIM for globalised manufacturing,* in *International Conference on Engineering and Technological Sciences* 2000: Beijing, China. pp. 319-327.

10. Dao, S.D., K. Abhary, and R. Marian, *Optimisation of resource scheduling in VCIM systems using genetic algorithm.* International Journal of Advanced Research in Artificial Intelligence, 2012. **1**(8): pp. 49-56.
11. Dao, S.D., K. Abhary, and R. Marian, *A stochastic production scheduling model for VCIM systems.* Intelligent Industrial Systems, 2016. **2**(1): pp. 85–101.
12. Dao, S.D., K. Abhary, and R. Marian, *An innovative model for resource scheduling in VCIM systems.* Operational Research: An International Journal, DOI: 10.1007/s12351-016-0252-y, 2016.
13. Dao, S.D., K. Abhary, and R. Marian, *An integrated production scheduling model for multi-product orders in VCIM systems.* International Journal of System Assurance Engineering and Management, DOI: 10.1007/s13198-016-0504-5 2016.
14. Liu, S. and M. Avery, *Alibaba: the inside story behind Jack Ma and the creation of the world's biggest online marketplace,* 2009, New York: HarperCollins e-books.
15. Waite, S., *Record-breaking Alibaba allocates to happy few*, in *FinanceAsia* (*Deal of the month: October*) 2014, Haymarket Media, Inc. pp. 8-9.
16. Dao, S.D., K. Abhary, and R. Marian, *Is a virtual computer-integrated manufacturing system feasible?*, *Lecture Notes in Engineering and Computer Science: Proceedings of The World Congress on Engineering 2016, WCE 2016*, 29 June - 1 July, 2016, London, U.K, pp. 733-736.
17. Lovász, L., *Review of the book by Alexander Schrijver: Combinatorial optimization: polyhedra and efficiency.* Operations Research Letters, 2005. **33**(4): pp. 437-440.
18. Dowlatshahi, M.B., H. Nezamabadi-pour, and M. Mashinchi, *A discrete gravitational search algorithm for solving combinatorial optimization problems.* Information Sciences, 2014. **258**: pp. 94-107.
19. Gao, L., *et al.*, *A hybrid algorithm based on a new neighborhood structure evaluation method for job shop scheduling problem.* Computers & Industrial Engineering, 2015. **88**: pp. 417-429.
20. Kuhpfahl, J. and C. Bierwirth, *A study on local search neighborhoods for the job shop scheduling problem with total weighted tardiness objective.* Computers & Operations Research, 2016. **66**: pp. 44-57.
21. López-Sánchez, A.D., *et al.*, *A multi-start algorithm for a balanced real-world Open Vehicle Routing Problem.* European Journal of Operational Research, 2014. **238**(1): pp. 104-113.
22. Salari, M., P. Toth, and A. Tramontani, *An ILP improvement procedure for the Open Vehicle Routing Problem.* Computers & Operations Research, 2010. **37**(12): pp. 2106-2120.

23. Kurdi, M., *A new hybrid island model genetic algorithm for job shop scheduling problem.* Computers & Industrial Engineering, 2015. **88**: pp. 273-283.
24. Cadoli, M., *et al.*, *NP-SPEC: an executable specification language for solving all problems in NP.* Computer Languages, 2000. **26**(2–4): pp. 165-195.
25. Shapiro, M. and E. Delgado-Eckert, *Finding the probability of infection in an SIR network is NP-Hard.* Mathematical Biosciences, 2012. **240**(2): pp. 77-84.
26. He, K., W. Huang, and Y. Jin, *An efficient deterministic heuristic for two-dimensional rectangular packing.* Computers & Operations Research, 2012. **39**(7): pp. 1355-1363.
27. Shahlaei, M., *et al.*, *Application of an expert system based on Genetic Algorithm–Adaptive Neuro-Fuzzy Inference System (GA–ANFIS) in QSAR of cathepsin K inhibitors.* Expert Systems with Applications, 2012. **39**(6): pp. 6182-6191.
28. Fahimnia, B., L. Luong, and R. Marian, *Optimization/simulation modeling of the integrated production-distribution plan: an innovative survey.* WSEAS Transactions on Business and Economics, 2008. **3**(5): pp. 52-65.

AN IMPROVEMENT OF HUMAN RELIABILITY ANALYSES THROUGH THE INCLUSION OF OPERATOR ANALYSIS

P. A. BAZIUK[†], J. NÚÑEZ MC LEOD and R. CALVO OLIVARES
CONICET/CEDIAC Institute, Eng. Faculty, Cuyo National University
Mendoza, Mendoza CO M5502JMA, Argentina
[†]pbaziuk@cediac.uncu.edu.ar
http://cediac.uncu.edu.ar/

S. S. RIVERA
CEDIAC Institute, Eng. Faculty, Cuyo National University
Mendoza, Mendoza CO M5502JMA, Argentina
srivera@cediac.uncu.edu.ar

A new dimension of analysis in human reliability models, operator analysis, and its possible mathematical quantification is hereby presented and discussed. Derived from psychometric studies, operator analysis is based on human abilities theory, and facilitates the quantification of the parameters required for a quantitative analysis of human reliability while also reducing the subjectivity in the analysis. Herein, the inclusion of cognitive, social and interpersonal skills puts focus on interpersonal differences and their impact on human performance. The innovative inclusion of Cattell-Horn-Carroll theory on cognitive abilities and Fleishman's premises on social and interpersonal skills simultaneously addresses several problems of existing human reliability models. Furthermore, it provides a theoretical basis for causal cognitive models of human error with a strong experimental support specifying their interrelationships and dependencies.

Keywords: cognitive abilities; human error; human performance; human reliability analysis; job analysis; operator analysis.

1. Introduction

Risk is currently a subject of significant interest to many areas of high social impact such as healthcare, nuclear power, aviation and military defense. Given that human error and the possibility of an incident are impossible to completely remove from complex technological systems, all potential interventions should be used to ensure that such aspects become, at least, controllable.

Risk management in these sectors represents the set of multiple complex actions which aim to improve the quality and ensure the safety of people and equipment. It is necessary indeed to consider error as an irremovable component of human nature. Therefore, given that it is impossible to completely eliminate human error and the associated risk, it is essential to promote appropriate working conditions. Yet and above all, it becomes crucial to perform a series of actions that make it difficult for humans to make mistakes and, thereafter, to establish defenses to reduce the mistake's consequences. Thus, it is necessary to understand the intrinsic mechanisms of human error in certain circumstances.

Human Reliability Analysis (HRA) studies seek to determine tendencies of persons to make mistakes in their professional performance. According to Kirwan[1] the three principal functions of HRA are "identifying what errors can occur (Human Error Identification), deciding how likely the errors are to occur (Human Error Quantification) and, if appropriate, enhancing human reliability by reducing this error likelihood (Human Error Reduction)".

HRA techniques or approaches can essentially be divided into two categories: first and second generation. Currently, we have come as far as to HRA dynamics and methods of a third generation, which are understood as an evolution of previous generations.[2]

First generation models focus on quantification without taken into consideration the cognitive causes of error. Based on statistical methodologies, these models make strong assumptions about human behavior, e.g. linear cognitive process that can be questioned.[3,4,5] Also, statistical methodologies fail to consider qualitative aspects, e.g. language parameters.[6] Complementarily and in order to improve such weaknesses, second generation models focus on human behavior and cognitive causes of human error. An important assumption, which can be strongly questioned, is that cognitive functions (perception, interpretation, planning and action) are separated.[3] In both generations, performance shaping factors (PSFs) are essential for producing quantifications. In first generation, PSFs are drawn from context or the environment while, for second generation, they are obtained from cognitive characteristics considering the aforementioned assumptions and without establishing a clear relationship between environmental and cognitive factors.[7]

Human behavior, according to both these methods, is assumed to be governed by two basic principles: the cyclical nature of human cognition and the dependence of cognitive processes on context and working environment. Within the scope of these models, cognitive functions (perception, interpretation, planning and action) are separately considered and individual differences are modeled as PSFs.

A lot of work has been done to make progress on the matter of the gap between cognitive psychology (human brain and human behavior knowledge and theories) and human reliability analysis. It is since such efforts that researchers have found an important relation between human error rate and reaction time and stress, the concept and theory of situation awareness,[8] workload and multiple resources in simultaneous tasks and the concept and theory of mental models.[9] Dekker and Hollnagel[10] carried out a strong critique to these models by establishing that these "folk" models are more descriptive than explicative and that the measures obtained from them reflect important aspects of operator situation but related with "intermediate cognitive states" rather than with real performance. The aforementioned authors proposed focusing on real performance instead of inferring uncertain states of mind.

In this paper, "operator analysis", a novel issue in HRA models based on human abilities model,[11] is discussed. The inclusion of the Carroll-Horn-Cattell (CHC)[12] theory of cognitive abilities based on psychometric tests should facilitate operator analysis quantification. In this line, possible mathematic techniques will be discussed.

2. Theoretical Basis: Human Abilities Model

Cognitive abilities taxonomy is a strong theoretical and experimental based model of human differences that favors objective quantification of such differences by means of psychometric tests. Measuring cognitive abilities is closer to measuring real performance than abstract constructions such as situation awareness or stress. Also, the later comparison between operator cognitive abilities and task demands results in human error possibility, which is in itself explanatory of underlying cognitive causes.

The theory of human abilities, including both cognitive abilities and social and interpersonal abilities, addresses the theoretical basis of HRA models:

(i) enables theoretical and empirical testability,
(i) captures most of current cognitive knowledge,
(ii) improves terms and construct nonspecificity.

2.1. *Human Cognitive Abilities*

For several authors, CHC theory, which gets its name from Raymond Cattell, John Horn and John Carroll, is the most comprehensive and empirically based psychometric theory. According to Carroll´s three-stratum theory,[13] relationships among individual differences can be classified into three different strata; that is, stratum 1 "narrow" abilities, stratum 2 "broad abilities" and stratum 3 "general abilities" or "g" factors. This theory is based on factor analysis. Stratum 1 is composed of first–order factors among tests and often referred to as primary mental abilities including nearly 80 abilities. Meanwhile, stratum 2 or second-order factors include 16 abilities while stratum 3 or third-order factors comprise general abilities or "g" factors (in spite of proving mathematically valid, no correlation with psychology constructs has been established and its validity has been questioned). There is structural evidence (differing in gender, level of education, ethnicity, nationality, language and historical period) confirming that primary abilities are part of second-factor high-order abilities positively correlated but independent (independence is demonstrated by way of structural evidence and distinct construct validity). Carroll (1993)[13] argued that these strata are more than mathematical entities as they can explain neurocognitive differences, i.e. broad abilities representing behavioral organizations founded in neural structures and functions. Furthermore, these factor scores are effective when accounting for behavioral differences.

Second-order abilities, or broad abilities, reflect most knowledge of the nature of human intelligence and initially were crystallized intelligence or acculturation knowledge (breadth and depth of knowledge of the language, concepts and information of the dominant culture), fluid reasoning or fluid intelligence (capacity to identify relationships, implication comprehension and inference drawing in novel or familiar situations), short-term apprehension and retrieval (also referring to short-term memory and working memory), fluency of retrieval from long-term memory, processing speed (rapid scanning and comparison of situations), visual processing (visual closure and constancy, object and pattern recognition), auditory processing (perception of sound patterns under distraction or distortion circumstances), quantitative knowledge (understanding and application of quantitative and numerical concepts, mathematical rules, numerical symbols) and reading and writing ability (includes basic reading and writing skills). A tenth ability, which is not currently assessed by tests drawn from a psychometric framework but can be interesting in terms of HRA domain, involves decision/reaction time/speed (reflecting the immediacy with which an individual can react to stimuli or a task). McGrew[11] included six additional broad abilities: general or domain specific knowledge (breadth, depth and mastery of knowledge in specialized subject matters or discipline domains that typically do not represent the general universal experiences of individuals in a culture and are developed through intensive systematic practice and training), tactile abilities, kinesthetic abilities, olfactory abilities, psychomotor abilities and psychomotor speed.

Interrelations among broad abilities do not point toward a single factor (factor g).[13] Focusing on their interrelations, second factor abilities can be classified as follows:

(1) vulnerable abilities which include fluid reasoning, processing speed and short-term memory (these abilities are likewise related to variables indicating neurological, genetic and aging effects),
(2) expertise abilities which comprise crystallized intelligence, long-term memory and quantitative knowledge (these abilities are related to learning and socialization; McGrew's domain specific knowledge ability should be included in this cluster),
(3) sensory-perceptual abilities, closely linked to strengths and weakness of sensory modalities, that present characteristics from both previous groups but do not clearly fit within any of them,
(4) psychomotor-physical abilities which were embraced by McGrew and exhibit different characteristics and evaluation tests drawn from the others abilities.

This classification should be relevant for HRA in terms of several aspects; namely, analyst training, initial human error possibility screening, task analysis, etc. Clusters 1 and 2 are usually grouped together into a more general "cognitive abilities" set.

The only antecedent relating human error to cognitive abilities was described by Buffardi *et al.*[14] Their approach was based on Fleishman's cognitive abilities taxonomy[15] derived from a correlation study of different jobs. We assume that the Cattell-Horn-Carroll classification theory is more appropriate due to the large extension and validation of their work, the fact that it is based on individual differences rather than differences between jobs and for being closely related to psychometric test that facilitate quantification. This last aspect, which is essential throughout this work, involves the fact that the problem in defining non-observable constructs for human reliability analysis is their subsequent measurement. The Cattell-Horn-Carroll theory tackles this issue.

2.2. *Teamwork: Social and Interpersonal Abilities*

Several authors have agreed on the importance of social and interpersonal abilities.[14,15,16,17,18] The analysis proposed focuses on the individual in team circumstances rather than on team level as a whole.[19] The abilities required to properly perform in a team may not be the same that are required for traditional individually oriented jobs.[16] Thereby, a fifth category of human abilities, or "social-interpersonal abilities", should be drawn upon in order to address human performance in teamwork. The level of social abilities relates to better adopting social roles required for managing conflict and handling work coordination.[16]

LePine and Van Dyne[20] followed the distinction regarding performance proposed by Borman and Motowidlo in 1993. Such notion indicates that there are two types of performances; that is, task performance (focused on activities that directly contribute to or support the transformation of inputs into outputs; i.e. operating machinery in a factory) and contextual performance (indirectly contributing with organizational success by maintaining or improving the organizational, social or psychological environment necessary for the technical core to function effectively and efficiently; i.e. helping and cooperating with others). LePine and Van Dyne studied 276 individuals and subsequently demonstrated that personality characteristics proved to relate more compellingly to contextual performance than to task performance.

Schumacher *et al.*[21] used a set of 19 social and interpersonal abilities extracted from F-JAS taxonomy[15]: achievement striving, agreeableness, assertiveness, behavior flexibility, coaching, coordination, dependability, negotiation, openness to experience, oral defense, oral fact finding, perseverance, resilience, resistance to premature judgment, self-control, sociability, social confidence, social conformity and social sensitivity.

Stevens and Campion[22] classified social and interpersonal abilities into two broad groups: interpersonal abilities and self-management abilities. Interpersonal abilities are then subdivided into three narrower categories: conflict resolution (manage conflict, air and relieve interpersonal friction, recognize types of conflicts and match a correct resolution strategy, negotiation or bargaining), collaborative problem solving (utilize the proper degree and type of participation, recognize obstacles and implement appropriate corrective actions) and communication (understand and utilize communication networks, communicate openly and supportively, listen non-evaluatively and appropriately use active listening techniques, maximize consonance between nonverbal and verbal messages, recognize and interpret the nonverbal messages). Self-management (of the team) abilities are divided into two categories: 1) goal setting and performance management (which help establish specific, challenging and accepted team goals as well as monitor, evaluate and provide feedback on both overall team performance and individual team member performance) and 2) planning and task coordination (plan and coordinate activities, information and task interdependencies among team members together with helping establish task and role expectations of individual team members and ensuring proper balancing of workload amongst the team).

As mentioned above, the problem in defining a set of abilities is their later measurement. Individual performance that facilitates effective team functioning is more related to contextual performance than to task performance.[16] Then, personality characteristics are likely to be good predictors of contextual performance which, in addition, is positively related to social skills. Therefore, it is possible to infer measurements of certain personality characteristics, contextual performances or social skills.

Burch and Anderson[23] indicated that personality should be a predictor of human performance and drew on the Five Factor Model personality taxonomy which classifies personalities into five groups: neuroticism, extraversion, openness to experience, agreeableness and conscientiousness. The most important factor included in this analysis is, in all probability, the relation that can be established between human performance and emotions such as anxiety, irritability, optimism or sense of responsibility. Results drawn from Barrick and Mount studies[24] revealed that one particular dimension of personality, conscientiousness, proves consistently correlated to all job performance criteria for all occupational groups.

Personality characteristics to be measured are conscientiousness, extraversion, agreeableness, emotional stability and openness to experience[21]. Such measurement entails the usage of the Personal Characteristics Inventory (PCI) method[15].

2.3. *Model Main Components Description*

The ideal HRA model has to quantify the HEP for probability safety analysis (PSA), to support the analysis post event for determine human causes of an accident, and to provide clues for system design, specially human-machine interface´s design.

To make that possible, the HRA model has four important components: task analysis, operator analysis, qualification method and quantification method (Figure 1). As summarized in Figure 1, the main results of human reliability analysis (system design, staff management, accident analysis, and HEP for PSA) drop on a continuum of quantitative and qualitative results: for staff management, qualitative results of HRA are more important, whereas for human error probabilities quantitative results are essential.

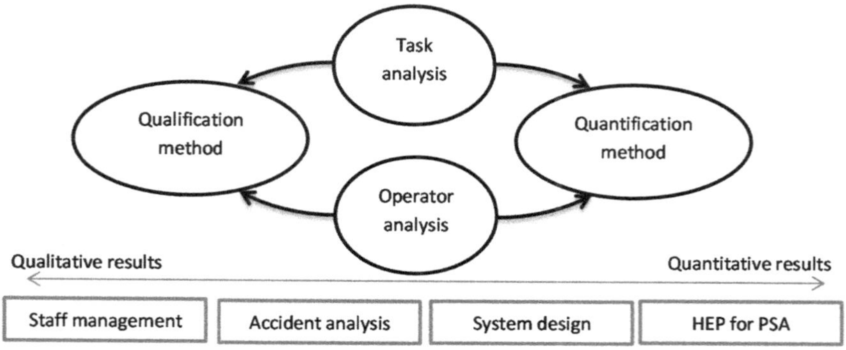

Fig. 1. Human operator in human reliability analysis: main component and results.

3. Towards Operator Analysis Quantification

Human abilities model inclusion in HRA involve operator analysis, this innovative step in HRA methodologies reduces uncertainty about individuals' actual behavior in a specific situation. Current HRA methods focus in task analysis, nevertheless some operator's characteristics, individual, and interpersonal differences are modeled by PSF.

HRA have a methodological module or component that performs human error probabilities calculation numerically, based on probabilistic techniques or others mathematical tools. These quantification methods are usually the main points of criticism of any model.[25] Operator analysis is based on a quantitative theory derived from psychometric studies therefore facilitates the parameters' quantification required in quantitative HRA.

Operator analysis is performed through human abilities assessment, measured with psychometric test. For example, fluid reasoning is measured by the following test: "complete sentences", "drawing analogies", "finding similarities", "arithmetic problems" and "numerical intelligence". These tests must be performed by a certified psychologist, and tests used must be validated for the region or country where the analysis is going to be done. Ideally, each operator must be submitted to all necessary tests in order to measure task relevant abilities. But in cases, where the operator that will perform the task is ignored, population means can be used.

Tests results are divers; they can be quantitative or qualitative. In the best of cases in terms of quantification, the numbers from quantitative results are in different ranges and scales. This requires the use of appropriate mathematical techniques. A bibliometric analysis[26] demonstrate that most recent and growing mathematical techniques in HRA are: fuzzy set, neural and bayesian networks (Figure 2).

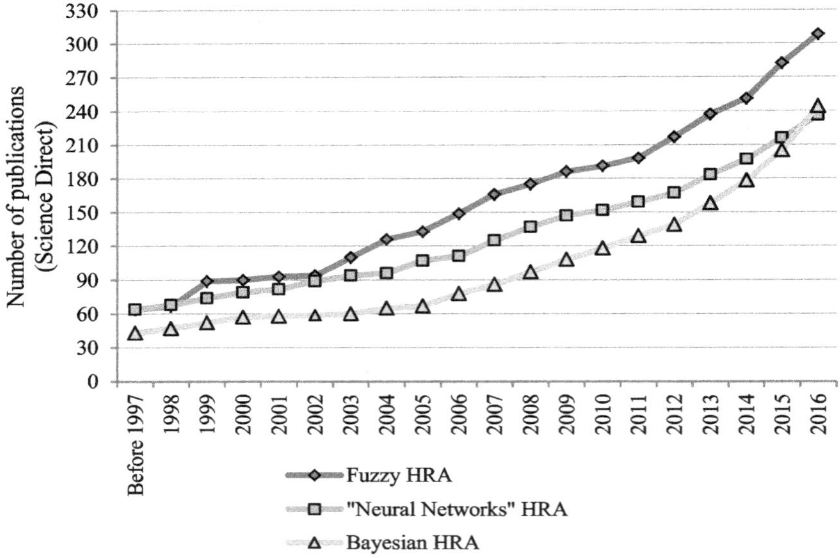

Fig. 2. Scientific publications evolution and trends about HRA quantifications techniques.

The most used technique is fuzzy mathematic, the most recent publications (2012-2015) were on[26]: uncertainty treatment (15%), expert judgment data treatment (45%), fuzzy fault trees (9%), performance shaping factors (18%), human behaviour model (13%). Fuzzy contributions improve HRA in five main aspects: (1) uncertainty treatment, (2) expert judgment data treatment, (3) fuzzy fault trees, (4) performance shaping factors and (5) human behaviour model. A recent and important line of research is about fuzzy-bayesian network (26% of fuzzy HRA applications from 2012 to 2015).[26]

From operator analysis and combined with an adequate cognitive task analysis, the degree of mismatch between the operator capabilities and task requirements can be estimated. This gap can be assessed in terms of human error, i.e. using linguistic variables modeled with fuzzy rules, error possibilities and probabilities may be calculated.

4. Human Abilities Model: Predictors, Correlations and Falsability

Dekker and Hollnagel[10] implied that phenomenon or construct explanations should be decomposed or reduced into fundamental elements that suggest possible measures favoring the corroboration of the given explanation. That is, falsifiability, an important principle in science.

Carroll's three-stratum theory provides a framework in which comparisons and correlations between psychometric variables and information processing variables can be made.[27] Also, correlations between human reliability constructs and cognitive abilities can be evaluated and would therefore prove the validity of measuring cognitive abilities instead of situation awareness, stress, workload, etc.

Some of these correlations have already been explored. For instance, a significant correlation between situation awareness and cognitive abilities was performed amongst military aviation pilots.[23] The prior example suggests that such type of correlation should be found in others cases and that, moreover, some other human reliability constructs should be equally supported by way of these means.

The validation (as opposite of falsifiability) criterion applied for the inclusion of cognitive abilities in human reliability analyses entails answering the question "are cognitive abilities predictors of human reliability?" to then perform evaluations on the bases of correlations.

McKenna, Duncan and Brown[28] observed a substantial correlation between typical intelligence tests and accident rates for bus drivers. Buffardi *et al.*[14] also perceived correlations between error rates and ability requirements for tasks in nuclear power plants.

Different abilities are required for different jobs and correlations between cognitive abilities and jobs should benefit task analysis, job redesign and operator recruitment and training. For example, Schumacher *et al.*[21] detected that 63% of the cognitive abilities present significant differences in requirement levels between analyzed jobs. Additionally, relevant divergences can also be established when examining social/interpersonal abilities.

The validity of personality characteristics as human performance predictors has also been assessed. While the highest validity is established for conscientiousness, there seems to be a "more complex pattern of relationships between personality and performance in jobs that involve interpersonal interactions than is captured solely by assessing Conscientiousness"[21]. Conscientiousness and emotional stability are positively correlated with job performance in the case of virtually all jobs.[20] These personality dimensions prove more stable in terms of validity for jobs requiring interpersonal components or teamwork organization (e.g. interactions with coworkers and customers). For example, sales and customer services exhibit the highest relations to conscientiousness of all while sales and managerial jobs also present such relations but in a lower degree with emotional stability and extraversion standing out. Another example involves customer services and how they are related to (in order of importance) conscientiousness, agreeableness, openness to experience and emotional stability.[21]

Another important issue concerning predictions of performance is the notion of incremental prediction. If the measurement of any given construct does not increment prediction accuracy or happens to reflect aspects that are related to other measurements, there is unnecessary redundancy. In other words, it is important that no construct measures can be inferred by means of other construct measures. In terms of Fleishman: "the fewest independent ability categories which might be most useful and meaningful in describing performance in a wide variety of tasks".[17] In the case of cognitive abilities, there is an experimental correlation between them that has led to the groundwork for the notion of general ability or "g" factor. There are two reasons for disregarding the usage of the "g" factor. Firstly and primarily, it is of dubious validity and, secondly, the use of different abilities facilitates task analysis for different jobs that may require different sets of abilities. In the case of social/interpersonal abilities, personality characteristics have been consistently found to be unrelated to cognitively oriented measures.

5. Conclusions and Discussion

The inclusion of underlying cognitive abilities in human reliability analyses involves relying on a theoretical base and, specifically, the Cattell-Horn-Carroll theory, which is largely experimental-based. This latter theory also facilitates consequent quantification since it is based on psychometric tests. Concurrently, the theory in question addresses the problem of strong dependence on expert judgment by drawing on these tests which are expressly developed to function as objective measurements.

The insight presented herein improves theoretical and empirical contrastability; incorporates deeper knowledge of cognitive sciences; sorts out the problem of lack of specificity in terminology and constructs in traditional models; reduces the structural model uncertainty, especially concerning the randomness of cognitive resources among individuals; and allows quantification by psychometric tests therefore decreasing bias introduced by expert judgment.

The innovation herein introduced would bring about the development of a model for which plausibility is based on the reviewed correlations. These correlations illustrate theoretical and conceptual contrastability and, thus, concluding that human abilities are a good predictor of human reliability and validating the use of human abilities over traditional model constructs (situational awareness, stress, complacency, workload, etc.).

The main contribution of this paper has to do with the inclusion of human abilities as the heart of reliability analyses and, consequently, the modeling of operator's individual specific abilities (including cognitive, physical and social ones) as a fundamental component of the system.

This contribution should also be utilized in human resources administration for tasks involving personnel selection, operators evaluation, training programs and training requirements.

A usual factor was omitted from this analysis: operating conditions. In unusual or, moreover, emergency situations, the human factor plays an important role as well. These types of operating conditions may increase task ability requirements while also modifying the ways in which people respond to stimulus. Emergency situations seem to encourage demand of social/interpersonal and sensory/perceptual abilities without modifying that of other cognitive abilities.

Acknowledgments

This research was supported by National University of Cuyo (Project Nº B033 - Res. N°3853/2016-CS) and the Argentine National Council for Scientific and Technical Research (CONICET, www.conicet.gov.ar).

References

1. B. Kirwan, *A guide to practical human reliability assessment* (Taylor & Francis, London, 1994).
2. V. Di Pasquale, R. Iannone, S. Miranda and S. Riemma, "An overview of human reliability analysis techniques in manufacturing operations," in *Operations Management*, Chapter 9 (2013).
3. W. J. Clancey, Situated Cognition: Stepping out of Representational Flatland, *AI Communications* **4**, 109-112 (1991).
4. A. Mosleh and Y. H. Chang, "Model-based human reliability analysis: prospects and requirements," *Reliability Engineering and System Safety* **83,** pp. 241-253 (2004).
5. E. Hollnagel, Human reliability assesment in context, *Nuclear Engineering and Technology* **37**, pp. 159-166 (2005).
6. L. V. Utkin and F. P. A. Coolen, "Imprecise reliability: an introductory overview," *Computational Intelligence in Reliability Engineering* **40**, pp. 261-306 (2007).
7. P. Baziuk, S. Rivera and J. Núñez McLeod, *Cognitive task analysis and Human Performance Shaping Factors. A reassessment from a second*

generation model, (American Society of Civil Engineers, 2013), pp. 1275-1281.

8. M. R. Endsley, "Situation awareness global assessment technique (SAGAT)," in *National Aerospace and Electronic Conference (NAECON)*, (Dayton, OH., 1988).
9. C. D. Wickens, "Multiple resources and mental workload," *Human Factors*, pp. 449-455 (2008).
10. S. Dekker and E. Hollnagel, "Human factors and folk models," *Cogn Tech Work*, pp. 79-86 (2004).
11. P.A. Baziuk, J. N. Mc Leod, R. Calvo and S. S. Rivera, "Modeling Human Reliability: the Underlying Cognitive Abilities," *Lecture Notes in Engineering and Computer Science: Proceedings of The World Congress on Engineering 2016*, 29 June-1 July, 2016, London, U.K., pp. 790-795
12. K. S. McGrew, "CHC theory and the human cognitive abilities project: Standing on the shoulders of the giants of psychometric intelligence research," *Intelligence* **37,** pp. 1-10 (2009).
13. J. B. Carroll, *Human cognitive abilities: A survey of factor-analytic studies* (Cambridge University Press, New York, 1993).
14. L. C. Buffardi, E. A. Fleishman, R. A. Morath and P. M. McCarthy, "Relationships between ability requirements and human errors in job tasks," *Journal of Applied Psychology*, pp. 551-564 (2000).
15. E. Fleishman and M. Railly, *Fleishman job analysis survey (F-JAS) Administrator's guide* (Management Research Institute, Potomac, MD, 1995).
16. D. P. Flanagan and P. L. *Harrison, Contemporary Intellectual Assessment: Theories, Tests, and Issues* (Guilford Press, 2012).
17. E. A. Fleishman and M. K. Quaintance, *Taxonomies of human performance: The description of human tasks* (Academic Press, Orlando, FL, 1984).
18. E. A. Fleishman, R. G. Kinkade and A. N. Chambers, *Development of a taxonomy of human performance: A review of the first year's progress (No. AIR-726-11/68-TPRI)* (American Institutes for Research, Pittsburgh, 1968).
19. F. P. Morgeson, M. H. Reider and M. A. Campion, "Selecting individual in team settings: the importance of social skills, personality characteristics, and teamwork knowledge," *Personnel Psychology*, pp. 583-611 (2005).
20. J. A. LePine and L. Van Dyne, "Voice and cooperative behavior as contrasting forms of contextual performance: evidence of differential relationships with big five personality characteristics and cognitive ability," *Journal of Applied Psychology*, 326 (2001).

21. S. Schumacher, M. Kleinmann and K. G. Melchers, "Job requirements for control room jobs in nuclear power plants," *Safety science*, pp. 394-405 (2011).
22. M. J. Stevens and M. A. Campion, "The knowledge, skill, and ability requirements for teamwork: Implications for human resource management," *Journal of management*, pp. 503-530 (1994).
23. G. J. Burch and N. Anderson, "Personality as a predictor of work-related behavior and performance: recent advances and directions for future research," in *International review of industrial and organizational psychology* (John Wiley & Sons, 2008).
24. M. R. Barrick and M. K. Mount, "The big five personality dimensions and job performance: a meta-analysis," *Personnel psychology*, pp. 1-26 (1991).
25. P. A. Baziuk, J. N. Mc Leod, R. Calvo and S. S. Rivera, "Principal Issues in Human Realiability Analysis," *Lecture Notes in Engineering and Computer Science: Proceedings of The World Congress on Engineering 2015, 1-3 July, 2015,* London, U.K., pp. 955-960.
26. P. A. Baziuk, S. S. Rivera and J. N. Mc Leod, "Fuzzy human reliability analysis: applications and contributions review," *Advances in Fuzzy Systems* **2016**, 9 Hindawi Publishing Corporation (2016).
27. V. C. Alfonso, D. P. Flanagan and S. Radwan, "The impact of the Cattell-Horn-Carroll theory on test development and interpretation of cognitive and academic abilities," in *Contemporary Intellectual Assessment: Theories, tests and issues* (2005), pp. 41-68.
28. F. P. McKenna, J. Duncan, I. D. Brown, "Cognitive abilities and safety on the road: a re-examination of individual differences in dichotic listening and search for embedded figures," *Ergonomics*, pp. 649-663 (1986).

MODELING SMART IMPELLER ACTUATED BY NITI SHAPE MEMORY ALLOYS*

K. O. SANUSI† and G. F. FUHNWI

The Faculty of Engineering and the Built Environment, Department of Mechanical Engineering Science, University of Johannesburg, Auckland Park Kingsway Campus, Johannesburg 2028, South Africa
†sanusik@gmail.com
www.uj.ac.za

O. L. AYODELE

The Faculty of Engineering, Department of Mechanical Engineering, Cape Peninsula University of Technology, Bellville Campus Cape Town, 1906, South Africa
ayodeleo@cput.ac.za

In this paper , analytical algorithms and advanced computer simulation package was used to study the performance of a pumping system and the best angle of attack for a Shape Memory Impeller by comparing the analytical algorithm and simulation. From the results , it shows that the best angle of attack is 12 degree at the outlet angle with respect to the inlet angle. Increasing the angle of attack from 35 degree to 45 degree at the outlet there is huge increase in flow rate by 63.47% and slight decrease in the impeller Torque from 35 degrees to 42 degrees by 0.72%. This study successfully conceptualised a Shape Memory Smart Centrifugal Impeller that will increase fluid flow rate through a given system if the fluid temperature increases above a given transition temperature using NiTi Shape Memory Alloys was achieved.

Keywords: Angle of attack; EFD; impeller; Nickel-Titanium. pumps; shape memory alloy; Smart.

1. Introduction

In recent years, convectional structures are being converged to smart structures by introducing smart material. Smart materials are normally composed of group 4-12 elements, also called transition elements. Compounds formed by transition elements use to have unique behaviour, they tends to behave strange way in response to external environmental changes, like pressure, temperature etc. [1]. Shape memory alloys (SMAs) are novel materials which have the ability to

*This work is supported by the National Research Foundation (NRF) and Faculty of Engineering and the Built Environment, Department of Mechanical Engineering Science, University of Johannesburg, Johannesburg, and Cape Peninsula University of Technology, Cape Town, South Africa.

return to a predetermined shape when heated [2, 3]. When this material is cold, or below its transformation temperature, it has a very low yield strength and can be deformed quite easily into any new shape which it will retain and it heated above its transformation temperature, it undergoes a change in crystal structure which causes it to return to its original shape [4, 5] (see Figs. 1 and 2). If SMA encounters any resistance during this transformation, it can generate extremely large forces. This phenomenon provides unique mechanisms for remote actuation [6]. Shape memory alloys such as Nickel-Titanium (NiTi) are distinguished by two unusual characteristics, i.e. the shape memory effect and pseudo-elasticity [7, 8]. Pseudo elasticity refers to the material's ability to be strained significantly and return to its unstrained configuration upon loading via a hysteresis loop [9–11]. A diffusionless transformation between two solid state metallurgical phases called Austenite and Martensite is responsible for both effects and can be induced by either changes in stress or temperature [10]. NiTi shape memory metal alloy can exist in a two different temperature-dependent crystal structures (phases) called martensite (lower temperature) and austenite (higher temperature or parent phase). Several properties of austenite NiTi and martensite NiTi are notably different [12].When martensite NiTi is heated, it begins to change into austenite (Fig. 3A). The temperature at which this phenomenon starts is called austenite start temperature (A_s). The temperature at which this transformation is complete is called austenite finish temperature (A_f). When austenite NiTi is cooled, it begins to transform into martensite. The temperature at which this transformation starts is called martensite start temperature (M_s). The temperature at which martensite is again completely reverted is called martensite finish temperature (M_f) [10, 13] . Composition and metallurgical treatments have dramatic impacts on the above transition temperatures. From the point of view of practical applications, NiTi can have three different forms: martensite (at temperatures below A_s), stress-induced martensite (at temperatures above A_f), and austenite (at temperatures above A_f). When the material is in its martensite form, it is soft and ductile and can be easily deformed. Super elastic NiTi is highly elastic (rubber-like), while austenitic NiTi is quite strong and hard (similar to titanium) (Fig. 3B).

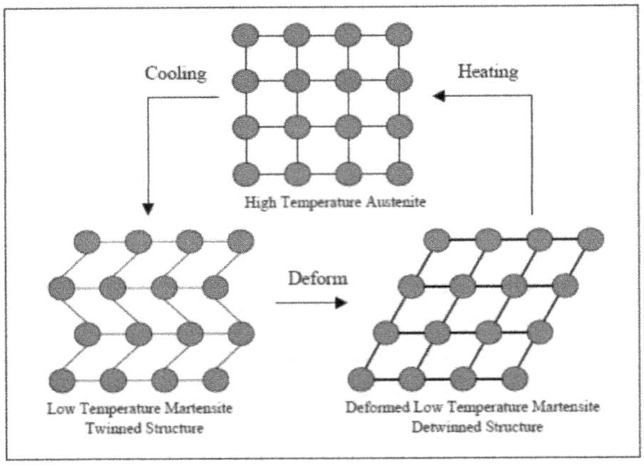

Fig. 1. SMA microstructural behaviour.

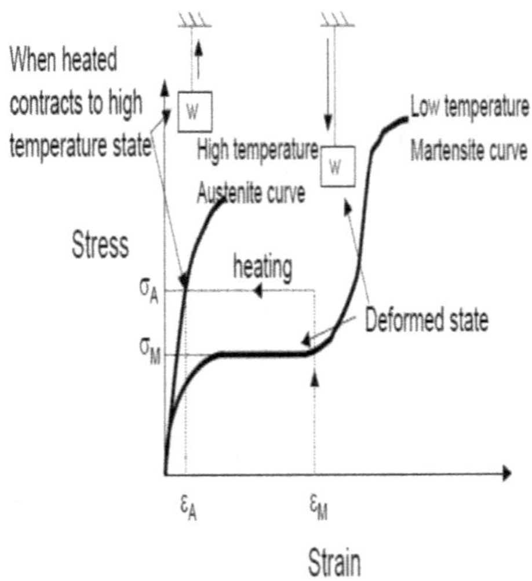

Fig. 2. Stress-Strain Memory effects of an SMA wire.

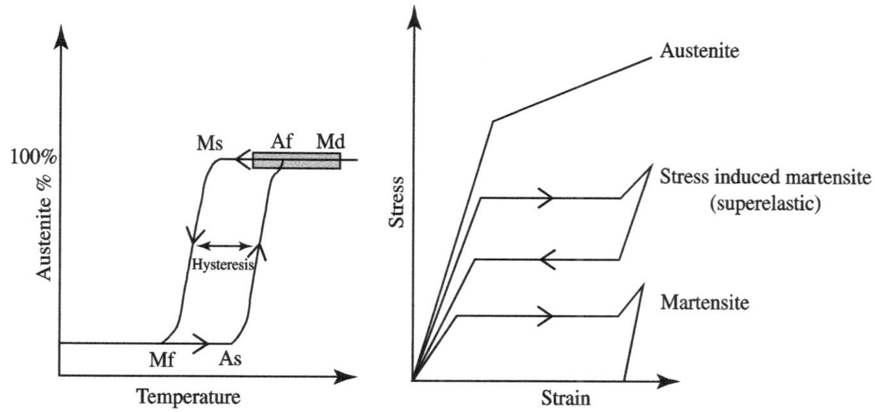

Fig. 3. (A) Martensitic transformation and hysteresis (= H) upon a change of temperature. As = austenite start, Af = austenite finish, Ms = martensite start, Mf = martensite finish and Md = Highest temperature to strain-induced martensite. Gray area = area of optimal super elasticity, and (B) stress-strain behaviour of different phases of NiTi at constant temperature.

A centrifugal pump is a device whose primary purpose is to produce pressure by accelerating fluid particles to high velocity providing them with velocity energy [ref 3] and one of the major parameters used to analyse the performance of a pump is the mass flow rate of fluid through the pump. For incompressible flow, it is more common to use Volume flow rate rather than mass flow rate [14, 15]. Manufacturers now develop a "smart" pump which incorporates micro-processors as part of their normal function [15]. A smart pumping system must be capable of knowing when to adjust itself to system changes without manual intervention. The system must also be fault tolerant.

The use of the one-way shape memory or superelastic property of NiTi for a specific application requires a piece of NiTi to be molded into the desired shape. The characteristic heat treatment is then done to set the specimen to its final shape. The heat treatment methods used to set shapes in both the shape memory and the superelastic forms of NiTi are similar. Adequate heat treatment parameters (temperature and suitable time) are needed to set the shape and the properties of the item [16]. They must usually be determined experimentally for the requirements of each desired part. Rapid cooling of some kind is preferred, such as water quenching or rapid fluid cooling. Certainly over the past few decades, significant progress has been made in the areas of pump hydraulics, mechanical design and applications through the use of computerized tools such as computational fluid dynamics (CFD) and finite element analysis (FEA). In this study, the shape memory effect exhibited by NiTi shape memory alloys will be utilized in the description of the Shape Memory Smart Impeller. It is thus believed that as the temperature of the fluid being pumped through the system is increased, the change in temperature will induce a phase transformation in the NiTi blade from Martensite to the parent phase, Austenite [18].

2. Constitutive Modeling of NiTi Shape Memory Alloys

The Brinson model uses constant material functions and offers the advantage of capturing the thermo-mechanical behavior of SMA's at any temperature. The model introduces the stress-induced Es and the temperature-induced Et martensite volume fraction. Based on the energy balance law, the constitutive relation for SMA's according to the Brinson model [45] is given as:

$$\sigma - \sigma^0 = E(\xi)\varepsilon - E(\xi^0)\,\epsilon^0 + \Omega(\xi)\xi - \Omega(\xi^0)\xi^0 + \Theta(T - T^0) \qquad (1)$$

where the quantities in the initial state is given by the suffix ^{0}, and the actual quantities without the suffix. E is the Young's modulus, and is given in terms of a linear function of the martensite phase as:

$$E = E_A + \xi(E_M + E_A) \qquad (2)$$

with E_A and E_M representing the 100% austenite and 100% martensite Young's Moduli respectively. ξ represent the martensite phase fraction present in the material and is composed of two types, i.e. thermally transformed and stress induced, given as:

$$\xi = \xi_T + \xi_T \qquad (3)$$

Ω represent transformation tensor, related to Young's Modulus, and is given in the form

$$\Omega(\xi) = -\varepsilon_{Tr} E(\xi) \qquad (4)$$

ε_{Tr} is the maximum transformation strain and for this particular situation assumed to be related to the displacement, δ, of the curved beam.

The final term in equation 1, $\Theta(T - T^0)$ is related to the thermal strain generated in the SMA due to the rise in temperature and Θ is equivalent to the thermal expansion of the SMA.

Assuming the material is in its initial state it is free from stress, in a fully twinned martensite state, and the elastic strain components are negligible compared to the transformation strain, equation 1 is now dependant only on thermal and transformation strains and is reduced to

$$\sigma = \Omega(\xi)\xi + \Theta(T - T^0) \qquad (5)$$

Assuming the transformation is not dependant on stress, Brinson [45] proposed the following relations associated ξ and T for equation (3).
For the conversion to Austenite

$$\xi_{M\to A} = \frac{\xi}{2}[cos[a_A(T - A_s)] + 1] \tag{6}$$

Where

$$a_A = \frac{\pi}{(A_f - A_s)} \tag{7}$$

When T > A_s
For the conversion to Martensite

$$\xi_{M\to A} = \frac{1-\xi}{2}[cos[a_A(T - A_s)] + 1] \tag{8}$$

where

$$a_M = \frac{\pi}{(M_s - M_f)}$$

when $M_f < T < M_s$ and $T < T^0$, else $\xi_{M\to A} = 0$
For transformation to austenite, the total transformation displacement δ is further given as

$$\delta = \delta_A + \xi_{A\_M}(\delta_M - \delta_A) \tag{9}$$

for transformation to martensite

$$\delta = \delta_A + \xi_{M\_A}(\delta_M - \delta_A) \tag{10}$$

With

$$\delta = \sqrt{(\delta_X)^2 + (\delta_y)^2} \tag{11}$$

δ_A and δ_M represents the transformation deformation of each phase. The components of displacement δ_X and δ_y represent the horizontal and vertical movement at the free end of the curved SMA insert.

3. Conceptualization and Structural Model of Shape Memory Smart Impeller

The NiTi strip is treated as a thin beam we find a possible solution to the operating mechanism of the Shape Memory Smart Impeller (see Fig. 4). At normal operating conditions, i.e. temperatures below A_s temperature, the NiTi thin beam assumes the inner shape of the impeller blade, The mechanism of actuation for our smart impeller is driven purely by thermal changes, i.e. increases in temperature from A_s to A_f will transform the twinned martensite to austenite, thus causing the shape change from a 30° inlet angle and 30° outlet angle to 30° inlet angle and 55° outlet angle. In the reverse transformation, i.e. a decrease in temperature from M_s to M_f, will transform the austenite back to twinned martensite and the SMA reverts back to the 30° inlet angle and 30°

outlet angles. Any change in the fluid temperature to above A_s and A_f will result in the NiTi SMA transforming from its low temperature martensitic state to the high temperature austenitic state. The latter state would be set to conform to the blade angle for 30° inlet angle and 55° outlet angle.

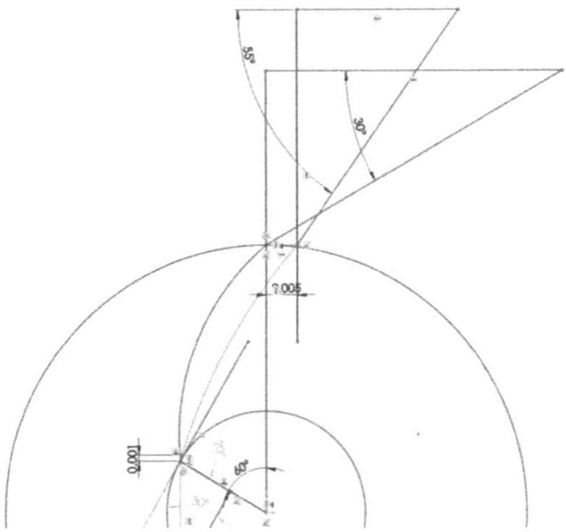

Fig. 4. Operating mechanism of the Shape Memory Smart Impeller.

Changing the centrifugal impeller blade outlet angle increases outlet velocity and in turn the pressure head, as well as the work done by the impeller on the fluid passing through the system. The impellers (see Figs. 5 and 6) were designed to have their outside blade sections remain identical while the inside blade section changes as the blade outlet angle is increased. The fluid flow passage (horizontal measurement) is decreased as the outlet blade angle is increased .For impeller blade outlet angles 45, 50, and 55 degrees the percentage decrease in the horizontal measure of fluid flow passage was determined as 1.086, 10.15, and 12.71 respectively. if we consider the entire blade or a section of it to be constituted of NiTi shape memory alloys and utilizing the shape memory effect it will be possible to change the blade outlet angle if the temperature of the fluid is increased above the NiTi specific A_s transition temperature, thus decreasing the fluid flow passage and increasing outlet velocity, pressure and work done by the impeller on the fluid. The NiTi SMA will be in its softer martensitic phase under normal condition. Any increase in the temperature of the fluid will then lead to the thermo-elastic phase transformation from martensite to austenite. The NiTi SMA should thus be trained to transform to the predefined shape of the 55 degree blade outlet angle on heating and revert back to the 30 degree outlet angle upon cooling.

Fig. 5. 3D Model of Centrifugal Pump Impeller showing blades, eye, shrouding disks.

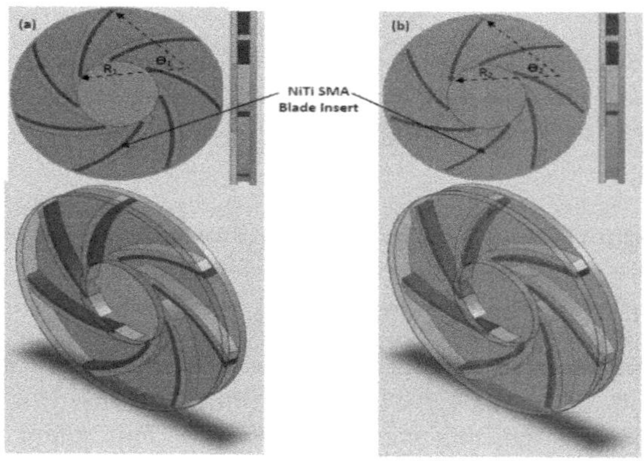

Fig. 6. Idealized model of the Shape Memory Smart Impeller with (a) showing low temperature configuration and (b) showing high temperature configuration.

Furthermore it assumed that the free end (see Fig. 4) moves within a guide until the final state is achieved. The shape memory effect exhibited by shape memory alloys will be utilized to affect actuation of the Shape Memory Smart Impeller. At normal operating conditions, i.e. fluid temperatures below austenite start temperature, the NiTi insert will conform to the shape of a blade with 30° inlet angle and 30° outlet angle. If the temperature of the water increases to above the austenite start temperature and continues to rise to austenite finish temperature thus fully transforming martensite to austenite, the NiTi insert will take the form of a blade with 30° inlet angle and 50° outlet angle. See Fig. 7, the transformation strain is thus equal to the displacement (δ) of the curved beam. This shape change due to the temperature increase causes an internal stress to be developed in the NiTi insert equal to the circumferential stress.

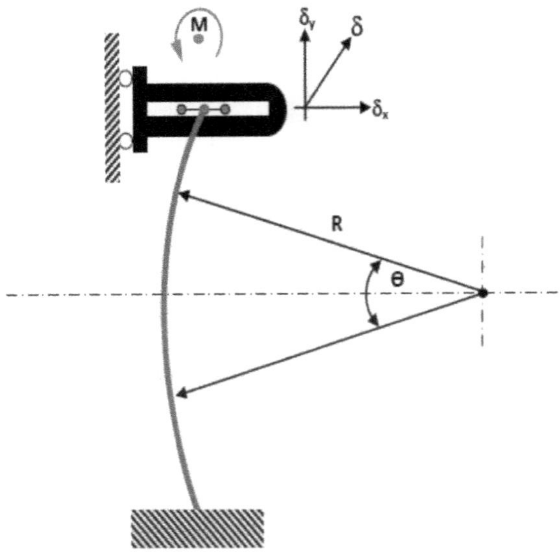

Fig. 7. Idealized model of operating mechanism of the Shape Memory Smart Impeller.

4. Simulation Algorithm for Engineering Fluid Dynamics

EFD.Lab, a full-featured general purpose CAD-embedded with finite volume fluid flow and heat transfer simulation tool was used for simulation, a rectangular adaptive meshing was used to mesh. This pump has a stationary axial inlet (an eye), a pipe section with a central body of circular arc contour, which turns the flow by 90 degrees from the axial direction. Water flows through a centrifugal pump having a rotating impeller (see Fig. 8). At the inlet's exit the radial water flow is sucked by a rotating impeller, which has seven untwisted constant-thickness backswept blades. Each blade is cambered from 30 degrees at the impeller inlet of 16 mm radius to range of 0 to 90 degrees at the impeller exit of 42 mm radius, both with respect to the radial direction. While keeping all configurations the same, only the impeller outlet angle was increase for each in impeller by 5 degree. The inlet angle for all impeller was 30 degrees. These blades are confined between the impeller shrouding disks rotating with the same (as the blades) angular speed of 2000 rpm. Volume flow rate of 0.3 m^3/s normal to the inner face of the Cover in the absolute frame of reference and Outlet Environment Pressure of 101325 Pa for the Environment pressure at the radial outlet face are used as the inlet and outlet boundary conditions respectively.

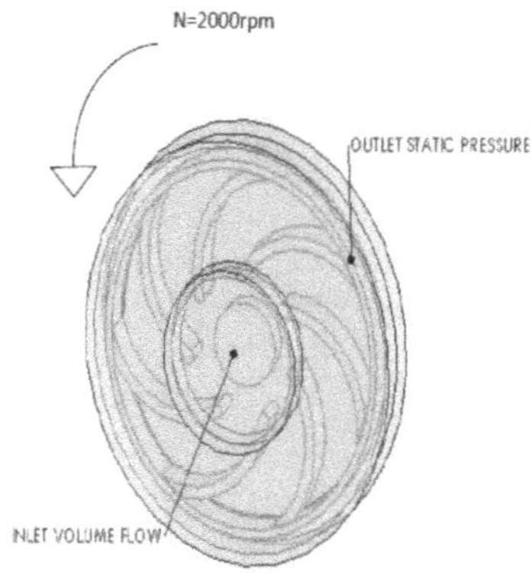

Fig. 8. Centrifugal pumps with rotating impeller.

5. Results and Discussions

Considering the centrifugal pump impeller rotating at 1500rpm with inside and outside diameters of 100mm and 200mm respectively and fluid enters the impeller radially. The impeller blade inlet and outlet angles are 200 and 300 respectively. The work-done by the impeller on the fluid per unit weight of fluid was 17, 23 Nm/N. vary the impeller outlet blade angle from 50 to 900 with increments of 50 and holding all other terms constant. The outlet blade angle versus relative inlet velocity (for constant inlet angle of 200) is showing in Fig. 10. The outlet blade angle versus relative outlet velocity (for constant inlet angle of 200) is showed in Fig. 11 and the outlet blade angle versus Work-done (for constant inlet angle of 200) is in Fig. 11. Form Figs. 9–11, it is clear shows that increases in the outlet angle (while keeping the inlet angle constant) will increase the relative inlet velocity, the relative outlet velocity, and Work-done on the fluid.

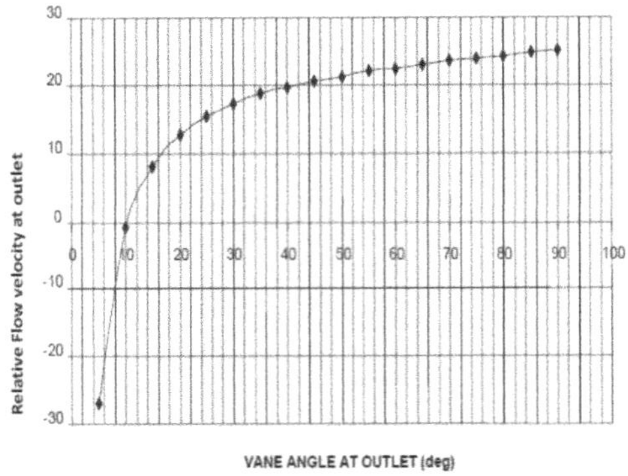

Fig. 9. A graph of Variable outlet angle versus relative flow velocity for a constant vane inlet angle of 20 degrees.

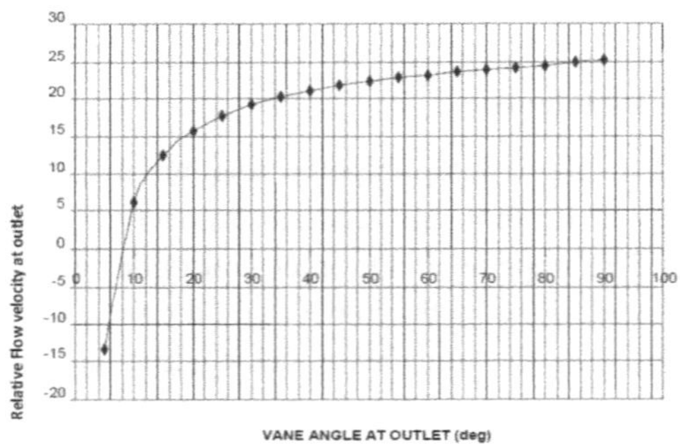

Fig. 10. A graph of Variable outlet angle versus relative velocity at outlet for a constant vane inlet angle of 15 degrees.

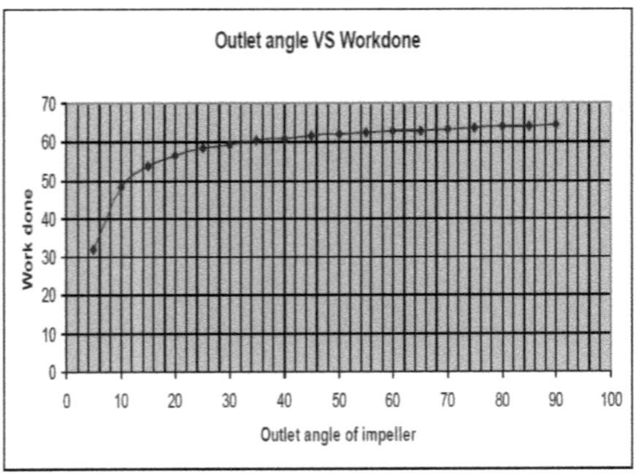

Fig. 11. A graphs of Variable outlet angle versus work done for a constant vane inlet angle of 5 degrees.

From Fig. 12 using the 3D Solid Modeling software package, Solidworks, to measure the middle of the inside blade to the center of rotation. For impeller blade outlet angles 45, 50, and 55 degrees the percentage decrease in the horizontal measure of fluid flow passage was determined as 1.086, 10.15, and 12.71 respectively. it is clear that the entire blade or a section of it to be constituted of NiTi shape memory alloys and utilizing the shape memory effect it will be possible to change the blade outlet angle if the temperature of the fluid is increased above the NiTi specific A_s transition temperature, thus decreasing the fluid flow passage and increasing outlet velocity, pressure and work done by the impeller on the fluid.

Considering the impeller, each having an inlet blade angle 30° but outlet angles varying from 30° to 55° with increments of 5°. the velocity vectors and static pressure distribution are shown in Fig. 13

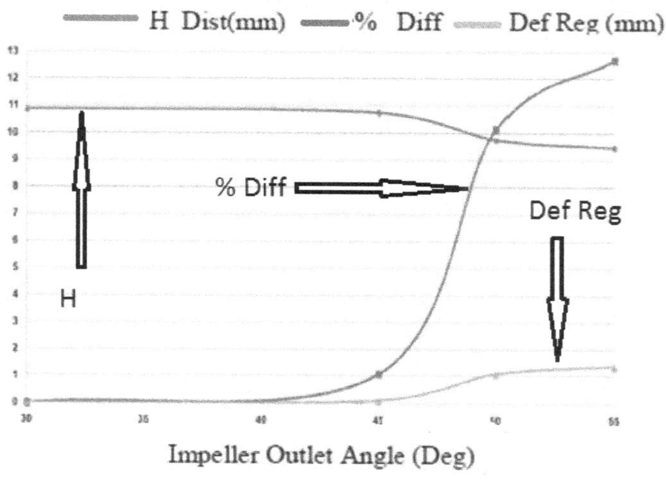

Fig. 12. Dimensional blade angle variations.

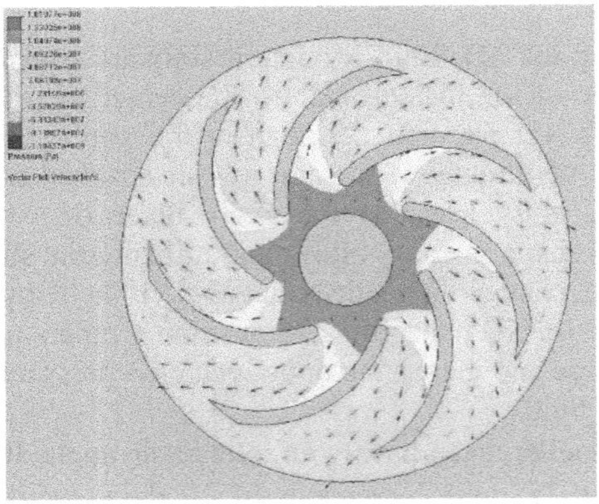

Fig. 13. Pressure plot.

From Fig. 14, it shows that the efficiency of the impeller drops drastically from 30 degrees inlet and 30 degrees outlet, but increases gradually as the outlet angle increase from this point. The graph also shows a characteristic optimum increase up to 45 degrees outlet where it starts to flatten out. This will allow us to have an angle of attack of 14 degrees from an inlet angle of 30 degrees. A Detail investigation reveals that there is a jump from 42 degrees to 45 degrees. Secondly there is a decrease from 45 degrees to 46 before increasing slowing onward. Investigating the torque on the system, it was revealed that despite an increase in the efficiency from 40 degrees to 42 degrees at the outlet, there is surprisingly a decrease in the torque in this section. See Figs. 15 and 16. A closer look at the torque alone on both sides of 42 degrees (where there is a drop) confirms our V-shape (see Fig. 16).

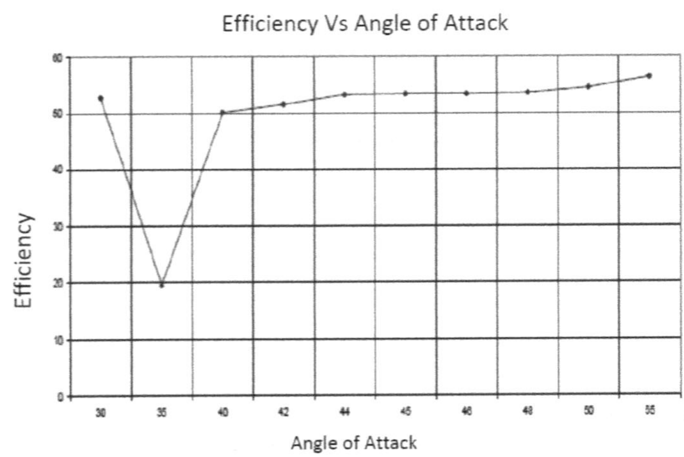

Fig. 14. Detail investigation of efficiency versus angle of attack.

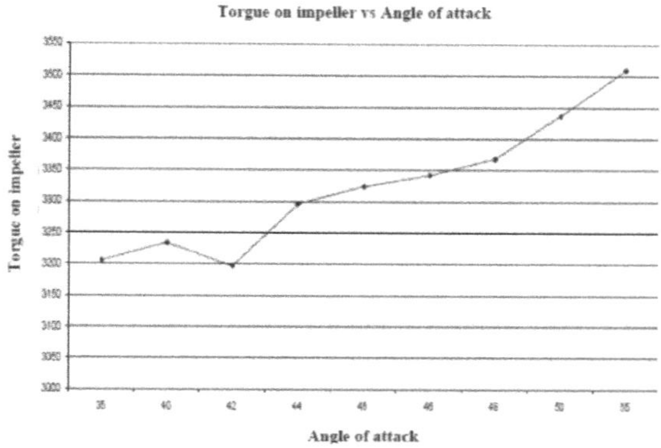

Fig. 15. Detail investigation of impeller torque versus angle of attack.

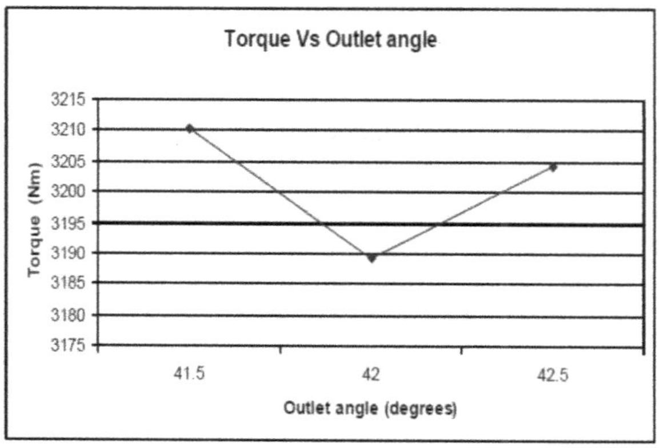

Fig. 16. Torque versus outlet angle.

6. Conclusion

This study successfully conceptualised a Shape Memory Smart Centrifugal Impeller that will increase fluid flow rate through a given system if the fluid temperature increases above a given transition temperature using NiTi Shape Memory Alloys was achieved .Based on the energy balance law, Brinson model was used to determine forward transformation, i.e. Martensite to Austenite, and the reverse transformation to describe the mechanism of operation of the Shape Memory Smart Impeller. The displacements change the outlet angle of the blade from 30^0 to 55^0.thus theoretically increasing the velocity at the outlet and increasing the head of the centrifugal impeller. For the computational fluid dynamics analyses, the results confirmed the optimum conditions of the specific system used to conceptualize the Shape Memory Smart Impeller. It also showed the formation of eddies, that would not typically be identified with the analytical solution. The benefits to for example the automotive industry would be the development of more efficient cooling systems. The operating benefits of such a pumping system include an increase in the cooling effect and higher efficiency thus providing less overheating, less impeller variation.

Acknowledgments

The authors acknowledged the National Research Foundation (NRF) and Faculty of Engineering and the Built Environment, Department of Mechanical Engineering Science, University of Johannesburg, Johannesburg, South Africa for the award of the research funds to support this study.

References

1. V. Msomi, G. Oliver, O. Philander and K. Sanusi, "A newly developed alternative way to describe shape memory alloy behaviour," in *International Conference on Computational Science and Its Applications*, Santander, Spain, 2011.
2. W.Ochonski, "Application of shape memory materials in fluid sealing technology," *Industrial Lubrication and Tribology,* pp. 99-110, 2010.
3. M. Braunovic, N. K. Myshkin and V. V. Konchits, "Fundamentals, Applications and Technology," in *Electrical Contacts*, CRC, 2006, pp. 565-583.
4. G. Song, B. Kelly, B. Agrawal, P. Lam and T. Srivatsan, "Application of shape memory alloy wire actuator for precision position control of a composite beam," *Journal of Materials Engineering and Performance,* vol. 9, no. 3, pp. 330-333, 2000.
5. G. Song, "Robust position regulation of a shape memory alloy wire actuator," *Journal of Intelligent Material Systems and Structures,* pp. 381-392, 2006.
6. Y. Sung, G. Shan and L. Shen, "Fracture toughening mechanism of shape memory alloys under mixed-mode loading due to martensite transformation," *International Journal of Solids and Structures,* vol. 38, pp. 4463-4476, 2000.

7. L. Petrini and F. Migliavacca, "Biomedical applications of shape memory alloys," *Journal of Metallurgy,* p. 15, 2011.
8. X. Wang, Y. Bellouard and J. Vlassak, "Laser annealing of amorphous NiTi shape memory alloy thin films to locally induce shape memory properties," *Acta Materialia,* pp. 4955–4961, 2005.
9. J. A. Shaw, "A thermomechanical model for a 1-D shape memory alloy," *International Journal of Solids and Structures,* pp. 1275–1305, 2002.
10. J. A. Shaw and S. Kyriakides, "On the nucleation and propagation of phase transformation fronts in a NiTi alloy," *Acta Materialia,* vol. 45, no. 2, pp. 673-700, 1997.
11. S. Nemat-Nasser and W. G. Guo, "Superelastic and cyclic response of NiTi SMA at various strain rates and temperatures," *Mechanics of Materials,* vol. 38, pp. 463–474, 2006.
12. G. Song, N. Maa and H.-N. Lib, "Applications of shape memory alloys in civil structures," *Engineering Structures,* vol. 28, pp. 1266–1274, 2006.
13. Y. Zhao, M. Taya, Y. Kang and A. Kawasaki, "A compression behavior of porous NiTi shape memory alloy," *Acta Materialia,* vol. 53, pp. 337–343, 2005.
14. S. Bansal, "Solid and fluid mechanics," New Delhi, Laxmi Publications, 2007, pp. 322-490.
15. P. E. Stavale, "Smart Pumping System: The Time Is Now," ITT Industries, Fluid Technology Corporation, Seneca Falls, NY USA, 2007.
16. K. Otsuka and X. Ren, "Physical metallurgy of Ti-Ni-based shape memory alloys," *Progress in Materials Science,* vol. 50, no. 5, pp. 511–678, 2005.
17. L. C. Brinson, "One-dimensional constitutive behavior of shape memory alloys: thermomechanical derivation with non-constant material functions and redefined martensite internal variable," *Journal of Intelligent Material Systems and Structures,* vol. 4, no. 2, pp. 229–242, 1993.
18. O. Kazeem Sanusi, Olukayode L. Ayodele, and G. Fuhnwi, "Computational Modeling of a Smart Impeller actuated by Shape Memory Alloys," Lecture Notes in Engineering and Computer Science: Proceedings of The World Congress on Engineering 2016, 29 June-1 July, 2016, London, UK, pp. 1032-1037.